# FIGHTING
# THE VIRUS
How Disease Modeling Can
Enhance Cybersecurity

# FIGHTING THE VIRUS

## How Disease Modeling Can Enhance Cybersecurity

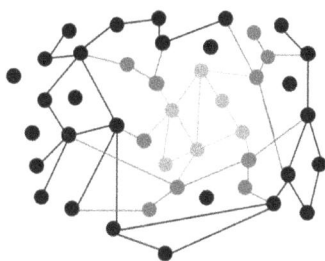

## Marc Mangel

University of Bergen, Norway

University of California, Santa Cruz, USA

**W🌐 World Scientific**

NEW JERSEY · LONDON · SINGAPORE · BEIJING · SHANGHAI · HONG KONG · TAIPEI · CHENNAI

*Published by*

World Scientific Publishing Co. Pte. Ltd.

5 Toh Tuck Link, Singapore 596224

*USA office:* 27 Warren Street, Suite 401-402, Hackensack, NJ 07601

*UK office:* 57 Shelton Street, Covent Garden, London WC2H 9HE

**Library of Congress Cataloging-in-Publication Data**

Names: Mangel, Marc author

Title: Fighting the virus : how disease modeling can enhance cybersecurity /
   Marc Mangel, University of Bergen, Norway & University of California, Santa Cruz, USA.

Description: New Jersey : World Scientific, [2025] | Includes bibliographical references and index.

Identifiers: LCCN 2025001889 | ISBN 9789811287565 hardcover |
   ISBN 9789811287572 ebook for institutions | ISBN 9789811287589 ebook for individuals

Subjects: LCSH: Computer security--Computer simulation | Diseases--Animal models

Classification: LCC QA76.9.A25 M31835 2025 | DDC 005.8--dc23/eng/20250404

LC record available at https://lccn.loc.gov/2025001889

**British Library Cataloguing-in-Publication Data**

A catalogue record for this book is available from the British Library.

For any available supplementary material, please visit
https://www.worldscientific.com/worldscibooks/10.1142/13710#t=suppl

Desk Editors: Aanand Jayaraman/Joy Quek

Typeset by Stallion Press
Email: enquiries@stallionpress.com

To the memory of William W. Hay, Donald Ludwig, and Colin W. Clark

# Foreword

Most generations face significant challenges a number of which result in significant levels of disaster. Some of these are human induced with conflict leading to wars being the most common. Others are induced naturally, pandemics and extreme geophysical and weather events being the most prominent.

However, it is arguable that the challenges faced by the current generation are at the extreme of such challenges. Clearly human conflict and the looming possibility of nuclear war is a massive issue, but I would also note that natural challenges driven by human demography and climate change pose extremely difficult issues of resource use for the future. Here the potential for science (including social science and engineering) to aid with prediction and mitigation is probably at its greatest ever. The use of science in this way goes back millennia. The key is that, properly used, science can enable complex problems to be addressed by suitable policy. Manifestly this is far from straightforward.

To return to my subject. This book deals with an important subset of the current challenges. The first is the potential for devastating epidemics of which the recent COVID pandemic provides a harsh reminder, but the second is a relatively new challenge of cyber security. The ubiquitous use of cyber systems in human activity is well known and their vulnerability is thus a constant concern.

In this book the author has used the observation that population biology and its application in dealing with epidemics has a similar mathematical structure to the apparatus needed for dealing with cyber security. The recent COVID pandemic has generated a wider understanding of the key epidemiological issues and how interventions can be used to mitigate the impact of the disease dynamics. This understanding is obviously varied within different societies, but key concepts appear to be part of most policy dialogue.

The mathematical apparatus for dealing with epidemiological problems has been developed over many years and continues to be developed in providing guidance to policy. The author then takes his fundamental observation and uses the mathematical apparatus developed in population biology to apply to cyber security problems. His preface develops these ideas in some detail and is a justification of the author's aim to provide understanding of the cyber domain to better aid policy.

Each of the individual chapters addresses a different and arguably more complicated issues of cyber security. I will not deal with them here as the author provides a summary of major insights in each chapter which are a helpful guide to the reader. There is a considerable amount of work in this book and the author's aim is to produce numerically literate readers who are capable of addressing key issues in the field. To this end he provides much detailed analysis and the potential to help with complicated problems that can only be addressed via computational methods.

A reasonable question is has he succeeded? I would argue that he has. It is a formidable and important work. I recommend it to you.

Professor Sir John Beddington CMG FRS
UK Government Chief Scientific Adviser 2008–2013
Emeritus Chair of Natural Resource Management
Oxford University

# Preface

## Welcome to the Journey

Cyber variability surrounds us – from problems with our personal electronics, to cyber crime in which cyber systems are held hostage, to threats on the critical infrastructure such as power or water plants being "taken down" by adversaries. We cannot both get away from these issues and live in the modern world.

This book is a contribution towards a theory of cyber variability, intended to inform the training of students and analysts (Dell Technologies, 2023) and provide a starting point for researchers. Maturing analytic capability for the study of cyber variability requires both a conceptual and modeling framework, a way for analysts to access this framework, and tests of the ideas developed in the framework. We (you, the reader, and I) will explore all of these. I also hope that people who are not modelers will read the book, conclude that cyber variability and its implications can be modeled, and that it is important to build groups that do such modeling.

When thinking about cyber variability there is an obvious and important role for systems engineering with a focus on specifics of particular systems, but there is also a role for the analysis of cyber operations at a level higher than the details of the specific operational systems. Often, the lack of a common analytical framework results in poor decision-making (Burris *et al.* 2010). Operations research/operational analysis (US/UK terminology) is a scientific approach to operational problems and will guide us in mapping from a top level conceptual framework to second level characteristics, metrics, and predictions.

**Cyberspace** is an odd domain (King and Gallagher 2020); indeed it is even difficult to define – we tackle that below – and I **boldface** it here because a definition is included both in the main text (below) and the glossary. If you are completely new to analytical thinking about cyber problems, I suggest that you take a look at Danzig (2014), Singer and Friedman (2014), and Libicki (2016). Danzig (2014) and King and Gallagher (2020) are available by download and broadly cover, in slightly different ways, the landscape of cyber security from an United States perspective. Libicki (2016) is a wonderfully informative introduction to cyber system variability, and is something to both read from cover to cover and have on your shelf (physical or electronic) to dip into from time to time. You will see that his ideas appear throughout this book, and we can start with advice given on page 1 that "when facing a problem such as the threat from cyberspace it pays to be serious but not desperate".

King and Gallagher (2020) provide a non-technical, sweeping, and well documented summary of some of the issues that we model in this book including

- A summary of the major public and private cyber threats;
- An overview of the recent major cyber operations publicly attributed to China (10 between 2006 and 2019), Russia (7 between 2007 and 2019), Iran (7 between 2011 and 2019), North Korea (5 between 2014 and 2019), and non-state actors (9 between 2011 and 2019);
- A strategy for layered cyber deterrence;
- The need to strengthen norms of international engagement and non-military tools;
- The need to promote national resilience (we will discuss resilience in detail later, but for now think of it as both the ability to fend off cyber attacks and recover quickly from attacks that are successful);
- Ways to reshape the cyber ecosystem towards greater security.

Furthermore, King and Gallagher (2020) identify major public- and private- sector cyber threats to the United States as attacks on elections and other democratic processes, espionage that undermines the military and defense industry, targeting civilian intelligence agencies, loss of leadership in research and development, cybercrime and ransomeware used for financial gain, theft of intellectual property, and putting critical civilian infrastructure at risk. Each of these is an important problem by itself. Zetter (2014) tells the story of Stuxnet, considered by many to be the first digital weapon, in an engaging and informative manner. It is also worth looking at both the section on cyber-related matters in the National Defense Authorization Act (NDAA; in the last few years this has been Title XV of the act and the most recent ones are easily found through a web search) and Arquilla and Ronfeldt (1997) who offered a view of this topic at the turn of the millennium – this book is still interesting, from both a historical perspective of more than 25 years and ideas that remain fresh.

**Cyber attack** broadly has three components (Libicki 2016, pp. 268–269): (i) generally understanding the vulnerabilities associated with the target, (ii) specifically searching for particular vulnerabilities and gaining a detailed understanding of the operations of the target, and once these are done, (iii) investing time and effort to conduct the attack and monitor its consequences. The target can recognize potential attack in (i) and (ii) and realized attack (iii) when it occurs.

We can broadly classify cyber security incidents as (i) Advanced Persistent Threat (APT, which we will call Simultaneous Cyber Operations) which is often cyber espionage, (ii) Distributed Denial of Service (DDOS), (iii) destructive cyber attacks, and iv) other cyber attacks, particularly cyber crime. Libicki (2016, Table 1.1, p. 6) lists 43 cyber attacks between 2005 and 2014 that consisted of nine APTs, six DDOSs, six destructive attacks, and 21 other attacks. His analysis gives a broad overview of the kinds of attacks, and discusses some detailed mechanisms, including an analogy from public health concerning botnets, in which users suffer little from being part of botnets but their victims can suffer a great deal; this is an argument for "herd immunity" in cyber space.

Perhaps the most famous example of a destructive cyber attack is Stuxnet (for more details see Chapter 1). However, in 2007

(3 years before Stuxnet), the Department of Homeland Security and Idaho National Laboratory (https://inl.gov/) conducted an experiment, the Aurora Generator Test, in which a generator received compromised instructions and went into self-destruct mode (Zetter 2014, Libicki 2016), showing that digital code could destroy a piece of physical equipment. At least 4 years of work getting makers of control systems to understand the vulnerabilities of their products preceded this test (Zetter 2014, p. 142). One important message of the Aurora test is that experimental tests of the ideas we are developing in this book are possible; indeed a major role of theory is to guide the experiments. For more details about the Aurora test, see Zetter (2014, pp. 160–164ff). The most important point here is that it is possible to conduct experiments on the effects of **cyber compromise**, subsequently we will discuss the possibility of experiments to measure the rates **co-compromise**.

### The Problems are Subtle

Regardless of your background, we are embarking an exciting and subtle adventure. Part of the subtlety is seeing how analysis, by which I mean mathematical and computational modeling, can interface with broader policy goals (e.g. White House 2023) that are rarely even stated quantitatively. When doing such work, I often re-read the instructions that Philip Morse, founder of the Anti-Submarine Warfare Operations Research Group (ASWORG) during World War Two (Morse 1977, Budiansky 2013), had read to members of ASWORG at least monthly "Our job is to help win the war, not to run it ourselves. We are novices at a task which has been worked at and thought about for many years. Our sole value is due to our specialized scientific ability. We begin to be useful when we can combine with our scientific training a practical background gained from contact with operating personnel. This practical background can only be obtained when the operating personnel trust us and like us" (Budiansky 2013, pp. 188–189). The oral history given by Phil DePoy (Sheldon 2016) provides an excellent overview of the ASWORG and its descendant, the Operations Evaluation Group at the Center for Naval Analyses, and a good overview of the operational perspective.

In order for our analytical models to change the way cyber security colleagues think about their work, we must gain their trust. Otherwise, as in the subtitle of Ambrus *et al.* (2014) we become irrelevant actors. Gupta (2014) visualizes a ladder for the way analysts and policy makers can interact, with the lowest rung being informal and unstructured interactions and the highest rung being formalized, centralized, continuous, and structured participatory involvement. Our goal is to climb this ladder in order to influence the way that people think about cyber security problems while maintaining analytical independence and creativity.

To do this, we will link population biology and cyber variability in a manner that does not require security clearance. But once you have developed the requisite skills, you might expect that your future work will be classified.

## Target Audience: This Book Might Be Written for You

My goal is to show the role of models and modeling in helping us thinking about and understanding the cyber domain, with the objective of providing input to decision-makers about how to play their operational hand.

Like my other books, this one is a mixture of research monograph and pedagogic text. That is, we will tackle new, unsolved, and important problems. In the course of doing this, I hope to expand your personal problem-solving kit (rather than just dazzle with a compendium of results), so that you are better prepared for tackling new problems.

I wrote this book with three main audiences in mind:

- *Population biologists.* If you are a population biologist, I hope that you will discover that broadening one's horizons about what it means to study a problem in "disease" opens a wealth of interesting and important questions and new opportunities for research.
- *Cyber security analysts.* If you already work on cyber security or cyber variability, I hope that you will discover new approaches to problems that interest you and an invigorated desire to learn some of the mathematical biology described in this book. If you pay attention to, rather than simply skip over, the equations you will

see that they are an embodiment of ideas and that the technical
details are really not that hard to master.

- *New operations analysts.* Operations analysis is a synthetic field,
  which is why, for example, the tradition in the Operations Evaluation Group of the Center for Naval Analyses was to encourage
  people to continue to have an interest in their disciplinary field
  (Mangel 1982, 2017). In general, simpler tools are better than complex ones and are more likely to deliver results that people need.
  But many new analysts come from an academic training where
  "harder is better" prevails. I hope that you will discover how powerful it is to use elementary mathematics in mature ways.

I hope that the book finds other readers too. Perhaps you are a hacker
who would like a framework for thinking about cyber compromise
and response to it; this is for you as well. Regardless of background,
I encourage you to read Thompson (2022) as a non-technical com-
panion to this book, but with a similar focus on models as tools for
understanding the world and communicating about it while at the
same time knowing the limits of models.

   If you are not at all familiar with disease modeling (although,
because of the coronavirus pandemic, I am confident that you are
familiar at least with the loose use of terminology, such as "grow
exponentially", "basic reproductive rate", or "host threshold density") that is no problem. In Chapter 1, we will explore the metaphor
linking disease modeling and cyber variability. Other places to start
learning about disease modeling are the short papers by Bjørnstad
*et al.* (2020a,b; 2000b) or Chapter 10 in Murray (2002).

   I expect that this is my last book (although I did so when I
wrote *The Theoretical Biologist's Toolbox* (Mangel 2006)), so I have
also included career advice and appreciate forbearance of readers for
that.

## Mathematical Analysis, Computation, and Potential Projects

We will understand the population biology of cyber systems using
mathematical analysis and numerical computation, and I will suggest

potential projects. The advantage of mathematical analysis is that we can demonstrate things that are true about the system we are considering. But often the system is too complicated for the mathematical analysis to be tractable, in which case we will use numerical computation to study its properties. In that case, one has to pick specific values of parameters and I will often suggest at the end of a section that you code a version of the model, confirm the results shown, and then explore what happens when parameters take different values. In particular, we expect some changes in the quantitative behavior as parameters change, but is there a change in the qualitative behavior? If yes, can you explain what is going on? In general, I will have conducted sensitivity analyses when selecting the particular results, and I will try to never mislead you.

Other times, I will suggest a **Potential Project**, which involves a re-conceptualization of some aspect of what we have done. Often these will be research topic for which I do not know what will happen. I will separate such topics in a box like this

> **Potential project**: Write a list now of what you hope to achieve by reading this book.

and hope that you follow through on some of these suggestions.

## Key Definitions and the Focus of This Book

The phrase "cyber security" evokes a wide range of interpretations (Sloan 2012, Chapter 6, Singer and Friedman 2014). For that reason, we now begin in earnest with some key definitions. Definitions are **boldfaced** at the time of introduction, and there is a glossary of terminology, symbols, and equations at the end of the book.

- A **cyber asset** is any kind of electronic information technology that may be operationally important in its own right or operationally important because of a linkage to other cyber assets or an **enabled physical system** whose performance depends upon the proper functioning of the cyber system.

- When cyber assets are linked they form a **cyber system** or **cyber infrastructure**. Stewart *et al.* (2010) offer two definitions of cyber infrastructure: "Cyberinfrastructure consists of computing systems, data storage systems, advanced instruments and data repositories, visualization environments, and people, all linked together by software and high performance networks to improve research productivity and enable breakthroughs not otherwise possible" and "Cyberinfrastructure consists of computational systems, data and information management, advanced instruments, visualization environments, and people, all linked together by software and advanced networks to improve scholarly productivity and enable knowledge breakthroughs and discoveries not otherwise possible." We will not be more specific than this, keeping with the idea that cyber systems are more easily recognized than defined (as Prisig (1999) notes about the word "quality").

- A cyber asset is **compromised** when its expected and actual behavior differ in significant ways. Note that this definition says nothing about *how* the compromise occurred. Because of natural variability, sometimes called noise, and operational factors cyber assets may wear out (Mangel and Brown 2022). The cyber asset is then compromised, and although this may also be an issue of cyber security it is not considered a cyber attack. Cyber **co-compromise** occurs when an uncompromised cyber asset has its behavior changed significantly after interacting with a previously compromised asset.

- A **cyber attack** occurs when compromise of a cyber asset is caused by an adversary. Cyber assets can be protected or hardened against attack, but few remain permanently invulnerable. When a cyber asset is compromised it may be repaired by humans or by other cyber assets.

- A **Cyber Protection Team (CPT)** is a trained group of experts who both maintain defense against attack and return compromised cyber assets to their functional state. Cyber assets successfully returned to operational status are **reset/restored**. We will use models to explore how one can structure the capabilities (e.g. the skills) and operations (e.g. timing of visits) of CPTs.

We thus interpret the terms cyber system, cyber infrastructure, and cyber space as broadly as possible. Valeriano and Maness (2015, pp. 23–24) provide a variety of definitions but settle on that of Nye (2011), which I paraphrase to include computers, networks, digital and cellular communication systems, fiber optic cables, and space based systems. Nye (2010, 2017) suggests that we conceptualize cyberspace as a hybrid of physical and virtual (informational) systems. The physical asset follows laws of physics and economics, and political jurisdiction. On the other hand, the informational asset is often embedded in a network with increasing returns to scale with more difficult jurisdiction. In addition, attacks on the informational asset have relatively small cost when compared to destruction of physical assets (which is one of the reasons that escalation from cyber to kinetic attacks, which we discuss in Chapter 3, is such a thorny topic).

There are many different mechanisms of cyber vulnerability and attack, including but not limited to: denial of service, direct attacks, phishing, reverse engineering, spoofing, and malware. Similarity, there are many different kinds of defensive counter measures, including but not limited to: security architecture, management of vulnerability, hardware protection, secured coding, digital hygiene, and protocol for responding to breaches. There is an enormous literature on the different mechanisms involved in cyber vulnerability, attack, and defense (Miller 2020, Microsoft 2022, Verizon 2022).

You will see that we do not need to model the specific mechanisms or cyber threats to understand the broader and higher consequences of cyber compromise. To focus at this higher level, we can conceive of cyber capabilities as consisting of cyber dependence (how much an actor depends upon cyber assets), the ability to conduct cyber offense, and the ability to conduct cyber defense in response to an attack. An example of such an assessment is given by Valeriano and Maness (2015, Table 2.1) for 10 countries ranging from Estonia and Israel to China, Russia, and the United States.

The use of mathematical models for understanding the broader and higher level consequences of disease has a long and rich history (Edelstein-Keshet 1988, Anderson and May 1991, Murray 2002, Keeling and Rouhani 2008). Our goal is to translate those ideas for an improved understanding of the variability and security of cyber systems.

## Things that are Helpful to Know

One of the powers of mathematics is that things which appear to be completely different on the surface are the same at a deeper level. I assume that you know or are motivated to learn:

- How to interpret a differential equation by understanding that $\frac{dx}{dt}$ denotes the rate of change of the variable $x(t)$ with time $t$. Of course, the more calculus you know, the better off you will be. Indeed, Richard Feynman told Herman Wouk when Wouk admitted that he did not know calculus "You'd better learn it. It's the language that God speaks" (Wouk 2010, p. 5). Long before Feynman, Galileo wrote in 1623 that "nature is a book written in the language of mathematics". Crease (2018, p. 68) wrote that "The most important lesson to be found in Galileo's image is the need to keep developing and revising the metaphors with which we speak about science."
    There are some instances in which I go into a bit more mathematical detail about a topic; such sections are denoted by $*$ and can be skipped if you wish.
- A few basic probability distributions such as the normal (Gaussian) distribution with mean $\mu$ and standard deviation $\sigma$, the binomial distribution that has its foundation in coin flipping, and the Poisson distribution in which a single parameter $\lambda$ characterizes the rate which events occur.
- Knowledge of a computer language. I have written the codes in R and implemented them in R Studio. Kadowaki (2023) is a nice starting point if you are unfamiliar with R but want to learn it. I have not posted code because if you want to really master the ideas here, you need to develop computer code yourself.

With the exception of R, the mathematical terminology will be clarified as we go along. If you are not familiar with these notions, keep in mind the comment of Crease (2018, p. 105): "When you are fully literate [in mathematics], nothing comes as a surprise. But mathematicians are made not born; in infancy they are not yet mathematicians and have to learn it – and in such learning often experience extensive transformations and reorganization of mathematical knowledge that they have only partially acquired."

## Acknowledgments

I wrote this book between July 2022 and November 2024. For the first year, my work was partially supported by the National Security Analysis Division (NSAD), Johns Hopkins University Applied Physics Laboratory (APL). I thank Phil DePoy, who reconnected Matt Schaffer and me after many years, and Matt for having invited me to think about the problems that colleagues at APL work on from the perspective of population biology and for his enthusiastic and constant intellectual support. This invitation was unexpected, lead to really interesting and important analysis, and bookends my career, which began doing national security analysis in the Operations Evaluation Group of the Center for Naval and Analyses (Mangel 1982).

Jimmie McEver and Alan Brown always made time – in person and on Zoom – for stimulating conversations and collaboration on research problems (Mangel and McEver 2021, Mangel and Brown 2022). I also thank Jimmie for suggesting that instead of a series of papers that I consider "a longer document or a book". Until he said that, I was pretty sure that I had no other books to write.

APL colleague Christine Fox motivated my thinking about multilateral cooperative cyber security agreements by asking me early my work with APL if there was anything to learn from the sustainable fisheries movement to encourage technology companies to cooperate in cyber defense. In Chapter 8 we will discuss what needs to been done to model the behavior of nations, rather than companies, but the ideas are transferrable.

I also thank APL colleagues Molly Gallagher, Toni Matheny, and Isaac Porche for discussions and questions that inspired various sections of the book.

I thank colleagues at the Theoretical Ecology Group, Department of Biology, University of Bergen for listening to seminars in which I developed these ideas, Katriona Shea and her group at Pennsylvania State University (especially Emily Howerton, Hidetoshi Inamine,

Joseph Keller, Emma van der Heide, Katie Yan), and Joe Travis for the opportunity to give practice talks as the ideas developed.

Mark McPeek's two brilliant books on community ecology (McPeek 2017, 2022) appeared during the time that I was doing this work. I thank Mark for a particularly great conversation in November 2022. Joe Travis has been a supporter of my research for many years and enthusiastic about the application of ideas from population biology to cyber systems. Joe also pointed out that this book is an example of translational research, something that the US National Science Foundation seeks to encourage.

I thank Spiro Stefanou for suggesting that I use broadband penetration and gross domestic product for the analysis in Chapter 8 on multilateral cyber security agreements.

I thank Jennifer Mangel for pointing me towards Pearson (2017).

I am very sorry that Dan Gillespie did not live to see this application of the Stochastic Simulation Algorithms, which I think would have both delighted him and made a full circle in our interactions that began more than 40 years ago (Gillespie and Mangel 1981).

Working with World Scientific Publications was a fantastic experience. Executive Editor Chris Davis approached me about this kind of book in October 2021; I responded that it was not clear that I had another book in me and asked him to contact me in a year. Almost to the same day in 2022 he wrote again; by this time Charles Fraccia had visited the Applied Physics Laboratory and the work in Chapter 2 was going very nicely. After that, Chris and I had regular communications as he provided key guidance. Joy Quek, Deputy Director (Life Science/Medicine/Chemistry Division) of the Editorial Department, also provided superb guidance and advice, as well as helped with some key issues of typesetting. Joy's colleague Rajesh Babu was able to help solve difficult LATEX problems from afar; it was great to work with him.

Were I a younger person, I would have taken another two years to expand many aspects of this book (making it bigger) but still consider it to be a finished product, albeit with various future directions that I leave for you to to explore. I thank Alan Brown, Bret Elderd, Edward Lerner, and Brian McCue for comments on the antepenultimate draft that improved both structure and writing, and caught some misprints. Some may remain, of course.

As with all of my books, I thank Susan Mangel for her patience with me when I fixated on something and for her probing questions about exactly what I was doing and why. For this book, I also thank Deb Denison for her patience during a visit in January 2023 when I was obsessed with working on it.

Marc Mangel, *December 2024, Tacoma WA*

# About the Author

Marc Mangel is Distinguished Professor of Mathematical Biology Emeritus at the University of California Santa Cruz and Professor of Biology Emeritus at the University of Bergen. He is fellow of the American Academy of Arts and Sciences, American Association for the Advancement of Science, Fulbright Foundation, John Simon Guggenheim Memorial Foundation, Royal Society of Edinburgh, and the Academies of Science of California and Washington State.

In June 2014, Mangel received Doctor of Science *honoris causa* from the University of Guelph. Citation in part: "You have profoundly influenced an entire generation of ecologists, environmental scientists and applied mathematicians on how to solve important practical problems and make the world a better place."

Mangel has numerous journal publications, about 37,500 citations in Google Scholar, a lifetime h-index of 89, and books that include *Dynamic Modeling in Behavioral Ecology*, *The Ecological Detective: Confronting Models with Data*, *Dynamic State Variable Models in Ecology: Methods and Applications*, and *The Theoretical Biologist's Toolbox*.

# Contents

# Summary of Major Insights

Each chapter ends with a non-technical summary of insights developed in the chapter. The full collection of insights follows here, to give you a sense of where we are going. There may be terms that you do not recognize, but they will appear in the appropriate chapter and be defined in the glossary as well.

### Chapter 1: Process Modeling in Population Biology and Cyber Systems

- The Resilience Stack provides a way for us to think about the hierarchical nature of cyber attack and defense.
- We cannot talk about redundancy unless we have in mind how the cyber system is to be used and how to evaluate its performance or the performance of the enabled physical system.
- The methods of population biology are natural tools for understanding the dynamics of compromise and variability in cyber systems. Three key ideas from population biology relevant to cyber systems are:

  (1) Populations (of organisms or cyber assets) consist of individuals with different characteristics and successful modeling of the dynamics of populations must have level of description that matches the question of interest.
  (2) Populations have dynamics on many different time scales, but often reach steady or quasi-steady states (which may include

periodic behavior [limit cycles]) in which dynamic processes
are balanced.
(3) Populations of organisms are governed by the fundamental law
of biology – evolution by natural selection – acting on expected
lifetime reproductive success of individuals as a proxy for the
long-term representation of genes in the population. Although
there is no similar fundamental law for cyber system, the
notion of fitness maps into the performance of a cyber system or the physical system that it enables.

## Chapter 2: The Pulse Attack Model (PAM)

- Following a pulse attack, compromise may persist even when the
  system reaches a quasi-steady state long after the attack has ended.
  Whether this happens or not is determined by co-compromise rate
  parameter in the cyber system and this is a property of the cyber
  network under the control of the defender.
- Consequently in anticipation of cyber attack, the very first pro-
  active defensive measure is to ensure that the co-compromise
  rate parameter is less than the threshold for persistence of co-
  compromise, which is a design parameter of the system.
- Cybersecurity drills can ensure that the co-compromise rate is
  below the threshold value by measuring the co-compromise rate
  parameter.
- There may also be design tradeoffs between the rate at which compromised cyber assets are discovered and removed to be reset and
  the rate at which they are returned to the uncompromised state.
  When this tradeoff exists, there is an optimal rate of removal to
  resetting that minimizes the maximum level of compromise (thus
  maximizing the minimum level of performance) but also a range of
  values of rate of removal to resetting that is consistent with making performance pretty good during the attack while eliminating
  compromise after the attack ends.

## Chapter 3: The Fundamental Model of Simultaneous Cyber Operations

- During persistent cyber operations, both sides will experience permanent degradation in which the steady state of uncompromised

cyber assets less than the initial number of uncompromised cyber assets is reached. This may correspond to a permanent degradation in performance of the cyber system or the enabled physical system, depending upon the parameters of the performance function and the number of uncompromised cyber assets in the steady state.

- The four-dimensional steady state of consisting of the uncompromised and compromised cyber assets of each side is unique and stable, meaning that regardless of the initial states of the cyber adversaries, the steady state will ultimately be approached. The approach to this steady state can include spiraling into it, rather than a monotonically approaching it.
- The value of the attack to the attacker can be determined from the probability of escalation by the adversary to a kinetic attack or a cyber attack on critical civilian infrastructure and the reduction in the performance of the adversary's cyber system or enabled physical system. These values also reach steady state values.
- Resilience, defined as the time to return to a near fully uncompromised cyber system if the cyber attack were to end, depends on both the rate at which compromised cyber assets are moved into the resetting pool and the rate at which they are moved from resetting to the uncompromised pool.
- It is a general property of simultaneous cyber operations that there are variables whose values shift from shrinking to growing or vice versa. An awareness of this property, as an anticipated and almost unavoidable phenomenon, is very important for decision-makers, who often pay great attention to trends but do not expect that they will reverse, or maybe even recognize that reversals are possible. Our analysis shows that a trend could shift direction without any outside influence.

## *Chapter 4: Beyond Determinism: Stochastic Versions of the PAM and FMSCO*

- The Stochastic Simulation Algorithm (SSA) allows us to move beyond the determinism implicit in ordinary differential equation (ODE) models of cyber systems. In particular, the SSA allows interpretation of the ODE models in terms of the conditional mean for an underlying stochastic birth and death process.

- A stochastic version of the PAM, operationalized by the SSA, produces distributions, rather than point estimates, of quantities such as the minimum number of uncompromised cyber assets, the time to recover to nearly all uncompromised cyber assets for the case in which the rate of co-compromise is below the threshold for persistence of compromise, or the minimum number of uncompromised cyber assets and the number of uncompromised cyber assets in the steady state for the case in which the rate of co-compromise exceeds the threshold for persistence of compromise.
- A stochastic version of the FMSCO, operationalized by the SSA, produces distributions for the steady state number of compromised and uncompromised cyber assets for each adversary and correlations between those values.
- In the stochastic version of the FMSCO knowledge by the X-side decision maker of the number of its uncompromised cyber assets says something about the number of uncompromised cyber assets held by the Y-side.
- Most importantly, for both the PAM and FMSCO using the SSA gives us a way to determine the range of variability due to fluctuations in the random processes underlying cyber attack and recovery. The SSA prevents us from misinterpreting variability as skill.

## *Chapter 5: Extensions of the Pulse Attack Model*

The PAM is intended as heuristic tool that has much in common with many systems but is not intended to model any specific system. The extensions of the PAM show the power of such an approach for developing understanding and quantitative predictions in cyber systems:

- The equations of the PAM generalize when there are multiple pulse attacks over time. In such a case, one needs to specify the times of the peak, dispersal parameter, and rate parameter of each of the attacks. Even when the rate of co-compromise is less than the threshold for the persistence of co-compromise, the cyber system may not recover fully and whether it does or not depends upon the timing and intensity of the pulses.
- By including the probability that the defender initiates either a kinetic attack or a cyber attack on critical civilian infrastructure,

the attacker can use the PAM to determine the attack rate parameter $a$ that is consistent with a targeted reduction in performance of the defender's cyber system or enabled physical system but below a threshold value for the probability that the defender responds with a kinetic attack or attack on critical civilian infrastructure. The PAM thus becomes a planning tool for the attacker.

- When a Cyber Protection Team (CPT) is required for restoration of compromised cyber assets, another time dependency is introduced into the equations for the PAM. Regardless of whether the CPT visits on a regular schedule or according to threshold number of compromised assets, the PAM generalizes directly, and should stimulate research by the defenders about the operation and effectiveness of CPTs.

- A straightforward extension of the PAM allows us to consider situations in which cyber assets that are restored to uncompromised status can be temporarily hardened to cyber attack, losing that defense over time. None of the qualitative conclusions based on the basic PAM, particularly concerning the role of the threshold level of co-compromise for the persistence of compromise in the steady state, change.

- A straightforward extension of the PAM allows us to consider situations in which assets are divided into those critical for the performance of the cyber system or the enabled physical system and those that have secondary roles, such as reducing the rate of attack on the critical cyber assets. In this case the number of differential equations in the model expands because we must track the dynamics of the two kinds of cyber assets. This extension allows us to study the role of protection of critical assets by secondary cyber assets. In particular, we can explore how performance of the cyber system or enabled physical system is shaped by the parameters characterizing the protection provided by secondary cyber assets and the rate at which critical cyber assets are returned to operational status.

- When cyber assets may be destroyed during an attack, the total number of cyber assets is no longer constant and we must make an assumption about the way destroyed cyber assets are replaced (or not). When destroyed cyber assets are not replaced so that the total number of cyber assets declines, the parameters of the performance function interact with the probability of destruction of an cyber asset during attack to determine how much steady state performance is degraded by the loss of cyber assets. When destroyed cyber assets are replaced from an on-hand pool of uncompromised cyber assets, the extension of the PAM allows us to determine the size of the reserve pool to maintain a sufficient level of performance. When destroyed cyber assets are replaced from an off-site pool of uncompromised cyber assets, a delay (the time for uncompromised replacement cyber assets to reach the cyber system) is introduced into the equations for the PAM. Such a delay can lead to oscillations into the dynamics of uncompromised cyber assets long after the pulse attack has ended.

- To design cyber systems that are Flexible, Adaptive, and Robust we can envision the parameters of the PAM, particularly the rates at which resetting cyber assets are returned to uncompromised states and at which compromised cyber assets are moved from the compromised pool to the resetting pool, as design parameters with a total resource constraint. Furthermore, we can add a third resource whose role is to reduce the rate of external attack. These considerations lead to a straightforward extension of the PAM. Although optimization of steady state dynamics or performance is clearly possible, sweeping over parameter values shows the optima for the number of cyber assets in the steady and performance and that a broad range of values that are close to optimum. That is, the surfaces characterizing the minimum and steady state levels of performance are relatively flat around the peak. The surfaces are broader when rate of co-compromise is less than the threshold for the persistence of co-compromise.

- Ultimately, one may choose to adapt the performance function in response to or anticipation of cyber attack. In this case the parameters of the performance function can be combined with those characterizing the dynamics of the cyber system to allow analysis of the tradeoff between the dynamics of the cyber system and performance of the cyber system or the enabled physical system.

### *Chapter 6: Extensions of the Fundamental Model of Simultaneous Cyber Operations*

- As with the PAM, we have seen how the FMSCO generalizes, which is what makes it a powerful starting point. That is, because the

FMSCO is not specific to any particular cyber system but has much in common with many cyber systems and with small modifications can capture other specific situations.

- When X-side cyber assets are used to hold compromise of the Y-side assets, performance of the X-side cyber system or enabled physical system may substantially decline. This is determined by the interaction of the midpoint of the X-side performance function and the number of X-side cyber assets needed to hold Y-side cyber assets in compromise. A similar conclusion applies for Y-side cyber assets holding X-side assets.

- Delays in the detection of compromise (moving compromised cyber assets from the compromised pool to the restoring pool) and restoration (returning cyber assets in the restoring pool to the uncompromised pool) may convert steady states that are approached monotonically to steady states that are spiral points. In such a case, one sides's uncompromised cyber assets may temporarily fall below a threshold for a kinetic response or an attack on critical civilian cyber systems. This important possibility needs to be clearly communicated to decision-makers.

### Chapter 7: Including a Distribution of Vulnerability in the Pulse Attack Model

- The rate of compromise can be modified to include vulnerability to attack by assuming that cyber assets with different levels of vulnerability are compromised at rates proportional to their vulnerability. The gamma density pinned down at 0 is a flexible means of capturing a distribution of vulnerability to compromise.

- The gamma density can be discretized into $N$ values of vulnerability and the PAM expanded to include $N$ equations for the dynamics of cyber assets with different vulnerabilities.

- When vulnerability has a discrete distribution, key decisions have to be made about (i) how co-compromise occurs and (ii) when cyber assets are reset, how the vulnerability of reset assets is determined. In this chapter, we assumed that there is a single pool of compromised assets and that co-compromise occurs at rate determined only by the rate of co-compromise, consistent with the assumption that the mechanism of external compromise and internal co-compromise are different. We investigated two choices for the vulnerability of reset cyber assets. In the first choice, reset cyber assets had vulnerability determined by the distribution of vulnerability at the time of the pulse attack. In the second choice, reset cyber assets return to the uncompromised pool with minimum vulnerability and then become more vulnerable as time goes on.

- When the co-compromise rate parameter is lower than the threshold for the persistence of compromise and cyber assets are returned proportional to the initial distribution of vulnerability, the overall number of uncompromised cyber assets and performance decline during the pulse attack but after the attack ends both return to the their values before the pulse attack. The numbers of cyber assets with different vulnerability decline during the attack and increase following the attack. However, because more vulnerable cyber assets decline at higher rates than less vulnerable assets, the distribution of vulnerability after the attack is different than before the attack and mean vulnerability may change from its value before the attack.

- When the co-compromise rate parameter is lower than the threshold for the persistence of compromise and cyber assets are returned with minimum vulnerability but then become more vulnerable as time progresses, during the pulse attack the overall numbers of uncompromised cyber assets and performance decline; after the attack ends both return to the their values before the pulse attack. In this case, there are transients in the number of cyber assets with different vulnerability, but ultimately all cyber assets have maximum vulnerability (unless other action is taken).

- When the co-compromise rate parameter is greater than the threshold for persistence of compromise and cyber assets are returned proportional to the initial distribution of vulnerability, during the pulse attack the overall numbers of uncompromised cyber assets and performance decline. After the attack ends compromise persists in the cyber system so that performance is permanently degraded. The numbers of cyber assets with different vulnerabilities decline during the attack and increase following the attack but not to their original levels. However, because more vulnerable cyber assets decline at higher rates than less vulnerable ones, the distribution of vulnerability after the attack is different than before the attack.

- When the co-compromise rate parameter is greater than the threshold for the persistence of compromise and cyber assets are returned with minimum vulnerability but then become more vulnerable as time progresses, during the pulse attack the overall numbers of uncompromised assets and performance decline. After the attack ends compromise persists in the cyber system so that performance is permanently degraded. The numbers of cyber assets with different vulnerabilities increase following the attack but not to their original numbers. Because more vulnerable cyber assets decline at higher rates than less vulnerable ones, the mixture of vulnerability after the attack is different than before the attack.

## Chapter 8: Bon Voyage: Future Directions

- Fruitful directions for future research include explicit models of the cyber system of the enabled physical system, incorporating human factors into the PAM and the FMSCO, and moving beyond dyadic interactions to consider multilateral cyber security agreements.
- Modeling the cyber system of the enabled physical system increases both the fidelity to operational situations and the complexity of the mathematical model but it will likely lead to new insights. Focus on a particular system, such as an electric grid and utility company is an appropriate starting point.
- Human factors are another natural extension of the PAM and FMSCO because humans are deeply involved in both creating compromise in cyber systems (e.g. by sharing thumb drives that carry malware) and detecting compromise (by recognizing anomalous situations). Modeling human factors will also expand the number of equations in the PAM and FMSCO.
- A Multilateral Cyber Security Agreement (MCSA) will help movement towards a global approach to shared cyber defense and response when cyber attacks occur. In this case a specific example of a performance function related to cyber systems is national Gross Domestic Product (GDP) as a function of national Broad Band Penetration (BBP), illustrated using nations that are member of the Organization for Economic Cooperation and Development (OECD). From it we learn:

(1) When the likelihood of cyber attack increases with the BBP of a nation and cyber resources can be allocated from BBP to unilateral defense, thereby reducing the number of successful attacks, there is a threshold level of BBP below which no defense is predicted (because adversaries will direct attacks to nations with larger BBP). After that, resources allocated to defense increase but at a decreasing rate, so that the fraction of BBP dedicated to defense is predicted to rise, level off, and fall for the largest values of BBP.

(2) Multilateral defense requires an assumption about to dependability of cyber security cooperation. When cooperation is guaranteed, nations are able to increase GDP relative to unilateral defense for all levels of BBP, with the largest relative gains going to nations with smaller BBP.

(3) When cooperation is not guaranteed, we must understand a nation's perception of how likely other nations are to deliver on their commitment to the cooperative security agreement. As a nation's perception of the trustworthiness of other nations declines, the gain from participating in a MCSA declines. Simulation methods allow us to assess the gain to a nation by participating in a MCSA and predict the fraction of nations joining the MSCA, illustrated with the OECD nations.

Chapter 1

# Population Biology and Process Modeling of Cyber Operations

*The important thing in science is not so much to obtain new facts as to discover new ways of thinking about them*

– Sir William Bragg

*But it's like an infectious disease, isn't it? Immunologists would have a field day researching the viral spread of compromising photographs on social media. I'd venture that the Spanish flu and Ebola combined couldn't touch the speed of photographic mortification spreading through cyberspace*

– Pearson (2017, p. 12)

## 1.1. Introduction

The broad idea underlying this book is that the development of cyber strategy (National Research Council 2010, Schneider 2020) will be improved by the development and use of **process based modeling**. With such models one describes the dynamics of the system of interest with explicit state (and often time) dependent functions of how the system changes (Pilowsky *et al.* 2022).

A cyber attack can be a single event or continuous, or something between those extremes. There are arguments that effective cyber strategies will involve Persistent Cyber Operations (PCO) (e.g. Goldman 2020, Nakasone 2020). That is "Continuous action in cyberspace for strategic effect has become the norm, and thus the command [USCYBERCOM] requires a new strategic concept"

(Nakasone 2020, p. 2). When multiple actors conduct PCOs, the result will be **Simultaneous Cyber Operations (SCO)**. Our (yours, the reader, and mine) goal is to contribute to the development of the required new strategic concepts and methods.

There are compelling arguments that a fruitful way to make predictions about the consequences of future cyber attacks is to conduct empirical inductive research (Gartze 2013, Maness and Valeriano 2016, Valeriano and Maness 2015, Rid 2012, Valeriano *et al.* 2018). Historical precedents for such work are the papers by Huth and Russett (1988) and Lebow and Stein (1990) concerning deterrence and the escalation of crises. Kreps and Schneider (2019) describe an experiment focused on the American public in which effect- and means-based routes to escalation in response to a cyber attack are explored. They found support for escalation through cyber thresholds; we will investigate this topic in Chapter 3.

Such inductive research in cyber incidents will allow us to replace worst-case speculation (Koppel 2015) by more nuanced understanding (Gomez and Whyte 2021). For example, the millions of daily intrusions into the US Department of Defense cyber systems are mainly routine scans from cyber criminals or nation-state adversaries intending to steal data rather than disrupt a cyber system (Lindsay and Gartzke 2018). There is now a sufficient body of data that one can analyze past cyber attacks and their consequences in order to make strong predictions about future attacks (in addition to the citations at the start of this paragraph, see Gross *et al.* (2018) and Warner (2020)). Valeriano and Maness (2015, p. 72) conclude that the "goal of theory is to generally explain the past, present, and future according to a set of foundations and ideas that guide interpretation and investigations. *With theory there must come predictions*" (italics added).

I fully concur, although we will develop a complementary approach to inductive, empirical study via deductive, process-based modeling. Process-based models allow us to reach beyond the empirical data, but be guided by it, and use deductive predictions to help guide data collection in the future (since the number of choices for what kind of data to collect is essentially unbounded). Put more simply, our goal is to show that cyber attack and its consequences can be modeled, in a way that theory conforms with experience

(Nakasone 2020, referring to Clausewitz) and leads to new insights and predictions.

## 1.2. The Disease Metaphor for Cyber Variability

In July 2018, shortly after I began consulting at APL, I made a trip there and spent a few days talking with many colleagues about the problems on which they worked. It was then that I met Jimmie McEver and as we sat together he drew a picture (Figure 1.1, upper panel) of the problems of cyber compromise that he was thinking about. I responded by drawing the analogue from the population biology of disease (Figure 1.1, lower panel). In that analogy there are susceptible (S), infected (I), and recovered individuals. When the link between recovered and susceptible individuals exists

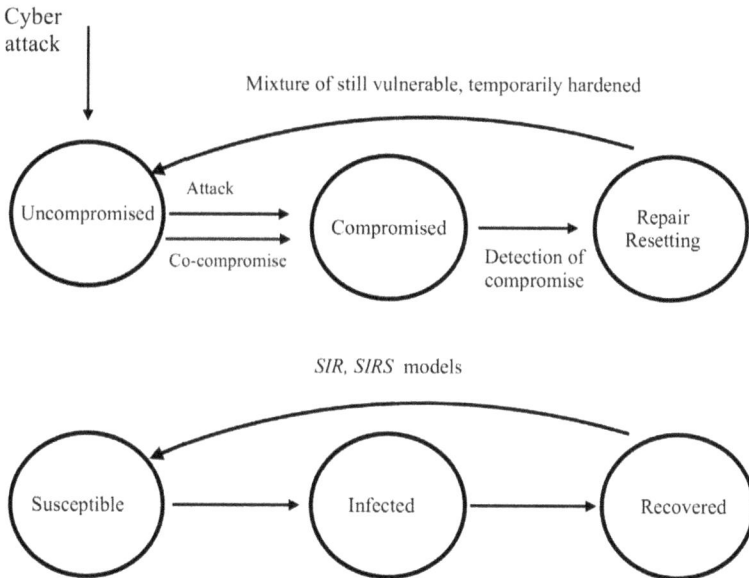

**Fig. 1.1.** Upper panel: A cleaned up version of a figure that Jimmie McEver drew on a white board at APL when we first met in July 2018. Lower panel: A cleaned up version of the figure that I drew after seeing his, explaining that there we many tools in population biology that can be used to make the metaphor between cyber systems and disease systems more precise.

because recovered individuals lose resistance to the disease organism, the model is the S[usceptible]I[infected]R[ecovered]S[usceptible] (**SIRS**) model; when recovery is permanent, the model is the S[usceptible]I[infected]R[ecovered](**SIR**) model. Such functional analogies from biology (Axelrod 2014), as well as other fields (Perkovich and Levite 2017), can lead to better understanding of the dynamics of compromise and recovery in cyber systems.

Terms from disease biology, including worm, reinfection, and outbreak appear throughout the cyber literature. It is even possible that mutations occur, as in the Conficker worm described by Zetter (2014, p. 54). Zetter also uses terms that include self-replicating (i.e. reproducing) worms (p. 58); inoculation (against self-infection) (p. 65), spreading mechanisms (p. 93) and writes that once Stuxnet spread in the world, it had "out-of-control spreading" (Zetter 2014, p. 357) and that copies of Stuxnet were "found in the wild" (Zetter 2014, p. 352). The Stuxnet virus also adapted to changes in the operating systems that it attacked (Bellovin *et al.* 2018, p. 270). Segal (2018 p. 326) also uses the word contagion.

As with disease organisms (and physical weapons), we can think of a payload and a delivery system. That is, a cyber weapon first has to reach the targeted cyber component (delivery), after which it may begin to compromise the target (payload). As in the natural world, there are many kinds of payloads and delivery systems. We will subsume these by the parameters characterizing the rates of attack, discovery of compromise, and return to the uncompromised state.

Clearly disease models are a metaphor for cyber compromise. Aristotle defined metaphor as a procedure in which the properties of one thing can be used to learn about the properties of another thing. Indeed, the Greek word *metaphora* translates to "to carry from one thing from another" (Reames 2024). Mathematical metaphors allow us to see that systems which appear to be very different actually have much in common with each other. The question is how far can we push the metaphor and what can we learn from it. Disease modeling is thus a lens (*sensu* Crease 2008, p. 178) that allows one kind of science (population biology) to approach another (cyber systems) in a focused manner. This focus allows the methods and thinking of first type of modeling to explore the second. Lafferty *et al.* (2008) discuss how models from the population biology of disease can be used to understand and confront the infectiousness of terrorist ideology. They too recognize disease models as a metaphor, and consider ways of making the metaphor practical, including goals for future research and remaining open questions (p. 204). Thompson (2022), in a book with a focus on models for climate, disease, and economics, has a superb chapter on models as metaphors.

### 1.2.1.   *On cyber and natural ecosystems*

The congruence between the visuals in Figure 1.1 is due the commonalities of the complexity in biological and cyber systems; Jimmie McEver had already been thinking hard about complexity in cyber systems for many years (Davis *et al.* 2000, McEver *et al.* 2019). Here's a modification of a table on page 204 of their paper, with corresponding thoughts about complexity in biological systems (Mangel *et al.* 1996, especially Appendix II, Levin 1998, 1999, 2003):

- Biological and cyber systems show *interdependence*, with the implications that cyber systems, like biological systems, are not easily understood by decomposing them into component parts because the relevant behaviors and vulnerabilities in one part or level of the system affect those in other parts of the system.
- Biological and cyber systems have many kinds of *nonlinearities* (in biology, "density dependent dynamics") with the implication that projecting forward from current conditions is more complicated than linear extrapolation, there may be multiple states of the systems, and rapid and unexpected transitions.
- Biological and cyber systems are *open* because new elements of the system are regularly introduced and/or current elements are regularly removed. Because forces outside of the system act on it, one must take care when focusing only on processes inside a preset boundary.
- Biological and cyber systems are characterized by a variety of *temporal and spatial scales* so that system properties and behaviors that are relevant for decision makers and users of cyber systems or managers of human interactions with biological systems exist simultaneously at multiple levels. The key to successful modeling of the dynamics of such systems is to find the level of description and the tools that match the questions of interest.

- Biological and cyber systems are embedded in *networks of interactions* so that causality is not always obvious, and in cyber systems responsibility and accountability are often not clearly definable.
- Biological and cyber systems show *emergence* in which novel system properties and behaviors appears at holistic levels from behaviors and interactions among the system elements. Such emergent properties and behaviors cannot be simply explained as the aggregate behavior of the system assets.
- The performance of biological systems is generally governed *evolution by natural selection* whereas cyber systems are governed by goals set by humans that are often vague, changing, unrealistic or conflicting.
- Biological and cyber systems show *adaptation and innovation* in which both the individual members of populations and the rules by which the system operates change, and intervention may stimulate adaptation (e.g. the evolution of resistance to pesticides in biological systems).
- Biological and cyber systems are *not fully transparent* so that there are often multiple hypotheses about why observed behaviors occur, and often insufficient evidence, time, or information to fully discriminate between them (Chamberlin 1897, Hilborn and Mangel 1997).

We now turn to modeling in population biology, and will construct equations for the dynamics of the SIRS model in the lower panel of Figure 1.1.

### 1.2.2.   *Modeling in population biology*

If you are coming from a background in physical sciences or engineering, you may expect that models require precise knowledge of the rules governing the dynamics of the system; if you are coming from statistical disease ecology, you may expect that the purpose of models is to fit data. Two of the giants of population biology in the 20th century, Robert May and John Maynard Smith (Figure 1.2) provided advice about this expectation in Graeme Farmelo's magnificent book on the great equations of modern science (Farmelo 2002).

**Fig. 1.2.** Two of the 20th century giants in ecological and evolutionary modeling, Robert May (left) and John Maynard Smith (right) moved to biology from physics and engineering respectively. We still have much to learn from them. Photo credits: Judith May and the University of Sussex respectively.

Maynard Smith (2002, pp. 196–197) noted that biological systems are inherently complex, but we tend to begin with simple models with just one or a few equations. He then wrote "How can simple equations help us in the face of all these complications?". The answer has a number of parts: (i) we isolate the phenomenon of interest, (ii) with a mixture of empirical work and intuition we hypothesize a mechanism driving the phenomenon, (iii) we test our guess by constructing equations that capture the proposed mechanism, solve (analytically or numerically) these equations and discover if "they generate the *kind* (italics in the original) of behaviour [British English in the original] that we observe". He concluded "In other words, we hope that our equations will predict qualitatively the right behaviour. Precise numerical fit is usually too much to hope for". Perhaps the biggest reason for qualitative predictions is that in any model we leave out a lot. But to make our models as complicated as the natural world itself makes them intractable. This begs the question as to why make mathematical models at all, rather than just sticking with verbal ones? Maynard Smith (2008, p. 211) addressed this point as well. First, constructing a mathematical model requires having an absolutely clear idea of what one assumes (even if it is a guess that ultimately gets modified). Second a good model will lead to testable predictions, and even qualitative predictions, such as the direction of change in the value of an output variable in response to a change in an input variable, allow empirical work to be guided by the models. And being able to use models to guide empirical work is a very valuable goal.

McElreath and Boyd (2007, p. 1) reinforce this idea when writing about mathematical models in the study of social evolution: "Mathematical models and the tools used to analyze them constitute the theoretician's laboratory. Simple mathematical models are experiments aimed at understanding the causal relationships that drive important natural phenomena ... These models are always too simple to make accurate predictions or even accurately represent how any real behavior evolves. Nonetheless, they have proven to be extremely valuable because they help us understand processes too complex to grasp by verbal reasoning alone."

May (2002, p. 2016) described his career path from astrophysics to theoretical ecology, writing that "It struck me that the equations that ecologists were using were in some important ways different from the more familiar ones of physics. The differences are not mainly in the technical nature of the equations, but rather that the equations of physics purport to give an exact account of whatever they are describing... [in the physical sciences] the more precise the information you put into the equations the more precise will be the equation's prediction. In population biology, things are often very different. There, the equations commonly refer to models of living systems that are always much too complicated to be amenable to the representational equations of the type beloved of physicists."

He then elaborated: "The models of biological communities tend rather to be of a very general, strategic kind – they are caricatures of reality. Just as a good caricature catches the essential truth behind the thing it is trying to depict but is forgivably vague about the unimportant details, so the most we can expect of the equations of population biology is that they capture the key points of the situation they are describing. So, for biologists studying animal populations, their equations are cartoons of reality, not the perfect mirror images sought by physicists. That's not to say that these biological equations are not vital to our understanding of nature. As the British mathematical biologist John Maynard Smith has noted, 'Mathematics without natural history is sterile, but natural history without mathematics is muddled.'"

When reviewing a book on mathematical principles of immunology and virology, Bangham and Asquith (2001) wrote: "It is a widespread fallacy that what mathematics contributes to biology is quantification of an otherwise innumerate science. But experimental biologists have long been expert at measuring and quantifying. *The real contribution of mathematics lies in a precise qualitative framework of reasoning. Experiment, however, is in no sense superior to theory, nor vice versa: both are necessary ingredients of a proper understanding of nature. An experiment done with no theoretical framework to analyze or interpret the results is meaningless; theory in the absence of experiment remains mere theory*" (italics added). Valeriano and Maness (2015, p. 46) concur: "Without theory, key aspects of cyber dynamics can be left unexplained, unexplored, or ignored. What processes are at work, why are they chosen, and how are they used? What predictions can be made, and what leads to the generation of these predictions in the first place?" If we view tactical outcomes in cyberspace as battle of wits rather than brawn using computer code rather than kinetic means (Lindsay and Gartzke 2018), tools that help sharpen thinking are extremely valuable. The techniques of the population biology of disease are such tools.

When applying the ideas from population biology, we will always have in mind the essential tension between the specific conclusions from a general model and the general conclusions from a specific model. By moving between these poles, we will learn much about cyber security from population biology.

Although a model may have technical aspects beyond the reach of some policymakers, lawyers, and judges, to be effective it needs to be presented in a manner that does not leave those individuals in a "modeling vacuum" (*sensu* Sulyok 2021, p. 292) where they simply have to trust that the assumptions and analysis of the model are sound. This void can be filled. For example, models are widely used in multilateral investment treaties and are used to (i) assess whether the evidence of a nation substantiates and risk and (ii) assess during a dispute whether a regulation was a reasonable response to an identified risk (Sulyok 2021, p. 215). Thompson (2022) discusses other ways we can bridge the gap between modelers and policy or decision-makers.

### 1.2.3.   *Characteristics of a good process model*

Our goal is to develop a process based model of a cyber system that can be used to explore the roles of attack and maintenance rates, detection capability, and other design parameters on the dynamics, particularly how compromise of the cyber system propagates, and effects the performance of the cyber system or the enabled physical system responds. Using the model will help identify what to measure to be able to assess vulnerability to cyber attack, the consequences of attack, and to identify design tradeoffs and routes to defense.

A good process model has at least these features

- A good process model should force one to think about the mechanisms underlying the system one is studying (Solow 2023). Thus, we should not expect for a model to "come off the shelf" and necessarily fit the system of interest. At the same time, a good process model should have much in common with many other cyber systems. One can then make changes to details of the model to make it particular to a specific cyber system. One of the (delightful) tensions of modeling in population biology is that these important points slightly contradict each other, and it is our job to figure out how to resolve this tension.
- A good process model should lead to unexpected predictions. If we are not sometimes surprised by the outcome of a model, then the predictions were built into it through the underlying assumptions.
- A good process model should guide us in the collection of empirical data.
- Especially for cyber systems, a good process model should be designed to help the development of strategic thinking, which some authors (Lin and Zegart 2019, p. 5) consider to be underdeveloped because so much specific information about cyber operations is classified (they note that in the early 2000s, the phrase "offensive cyber operations" was classified). However, even though many military cyber operations are classified, non-military users of cyber systems are frequently discovering, attributing, and mitigating cyber compromise. Doing so brings more data into the public domain. We should expect a multiplicity of ways to achieve persistence in the cyber domain (Smeets and Lin 2018) and a good process model will help identify the routes.

- A good process model should provide a quantitative framework for existing qualitative descriptions for the behavior of cyber systems (Smeets and Lin 2018).

### 1.2.4. *The equations of the SIR and SIRS models*

With these ideas in mind, let us construct a dynamical model for the SIR and SIRS models in Figure 1.1. If you are already a quantitative population biologist, you will find this section easy going.

#### 1.2.4.1. *Symbology*

To begin, we need symbols for the three pools of individuals in Figure 1.1. Letting $t$ denote time, it makes sense to let $S(t), I(t)$, and $R(t)$ denote the number of Susceptible, Infected, and Recovered individuals at time $t$, and adopting one of the most common notations from calculus, we denote their rates of change by $\frac{dS}{dt}, \frac{dI}{dt}$, and $\frac{dR}{dt}$ respectively. Next, we need assumptions about how these rates of change are determined and make the following assumptions:

- Susceptible individuals become infected at a rate proportional to the product of the number of susceptible individuals and the number of infected individuals with proportionality constant $\beta$, i.e. $\beta S(t)I(t)$.
- Infected individuals move from the infected pool to the recovery pool at a rate proportional to their numbers with proportionality constant $\mu$, i.e. $\mu I(t)$.
- Individuals recover at a rate proportional to the size of the recovering pool, with proportionality constant $\gamma$, i.e. $\gamma R(t)$. These individuals join the susceptible pool in the SIRS model ($\gamma > 0$), or remain in the recovery pool in the SIR model ($\gamma = 0$).

#### 1.2.4.2. *The dynamic equations*

Susceptible individuals who become infected move from the susceptible to infected pool; infected individuals who start recovery move from the infected to recovery pool; and when individuals lose immunity (the SIRS model), recovered individuals losing immunity move to the susceptible pool. Thus, the rates of change of the three kinds of individuals are

$$\frac{dS}{dt} = -\beta I(t)S(t) + \gamma R(t)$$

$$\frac{dI}{dt} = \beta I(t)S(t) - \mu I(t)$$

$$\frac{dR}{dt} = \mu I(t) - \gamma R(t)$$

When we add these three equations, the sum of the right sides is zero. That is, the rate of change of the sum of the three pools is zero, which means the population has a constant size (there are no births, deaths, immigration, or emigration or that all of those processes somehow balance each other). If we let $N$ denote the constant size, then $N = S(t) + I(t) + R(t)$. The assumption of constant total size is handy because it allows us to reduce the three equations to two by writing $R(t) = N - S(t) - I(t)$; we will exploit this property in the next chapter.

In general, we will adopt the convention of not writing the time dependence of dynamical variables on the right side of equations.

### 1.2.4.3.   *Always confirm understanding by checking units*

One of the fundamental principles of applied mathematical modeling is that things in the actual world have units and the units on the two sides of an equation have to match. Let us employ this principle to determine the units of $\beta, \gamma$ and $\mu$, using $[\bullet]$ to mean the units of whatever is inside the brackets. We also employ the fiction that $\frac{dS}{dt}$ is just a fraction (it is much more special than that), so that the numerator has units of individuals and the denominator has units of time. We conclude $[\frac{dS}{dt}] = \frac{\text{susceptible individuals}}{\text{time}}$; the same will be true for the units of $\frac{dI}{dt}$ and $\frac{dR}{dt}$ with change of the adjective to infected or recovering. The right side of each equation needs to have units that match.

First consider the term $\beta SI$ (where I am dropping the notation of time dependence). This has to have units $\frac{\text{susceptible individuals}}{\text{time}}$ and since $I$ has units of infected individuals, $\beta$ must have the admittedly weird units $1/(\text{infected individuals} \cdot \text{time})$. You will sometimes see people replace $\beta SI$ by $\frac{\beta'}{N} SI$, making $\beta'$ a pure rate with units $1/\text{time}$.

Reasoning as follows for the other parameters, we conclude that $\gamma$ and $\mu$ are rates with units of $1/\text{time}$.

### 1.2.4.4. *Starting population sizes*

In order to describe how the population changes, we need to give starting population sizes, which are called initial conditions. By convention, the starting time is usually $t = 0$, so that we specify $S(0) = S_0$ and $I(0) = I_0$, from which $R(0) = N - S_0 - I_0$. One natural starting question is "what happens to a population of susceptible individuals when an infection is introduced?" and we might think to set $S_0 = N, I_0 = 0$ and $R_0 = 0$. But if $I = 0$, then the right side of these equations tells us the population will not change. This is a problem, which we often solve by modifying the question to "what happens if we introduce one infected individual into a population of otherwise susceptible individuals?" and thus set $S_0 = N - 1$ and $I_0 = 1$. This kind of reasoning leads to the calculation of the number of new cases caused by a single infected individual entering an otherwise susceptible population (this is called the basic reproductive rate $R_0$ of a disease).

The broader idea is that there somehow has to be an "injection" of infected individuals into the susceptible population. For cyber systems, we do this in Chapter 2 with a single pulse of attack or in Chapter 3 with continuous attack.

### 1.2.4.5. *Metaphor again: On individuals and populations*

None of the equations we just discussed focus on a particular individual but on populations of individuals, and they is why they are so powerful. When an individual is sick, depending on the severity of the illness they could rest, take over the counter medication, or seek medical attention. But if we want to understand how an illness will spread in a population, we have to focus on the interactions between the disease agent and the individuals in the population, and on the interaction between healthy and sick individuals in the population. In a similar way, when one's computer is having problems, depending on the severity of the problem one might turn it off and back on, run program to defeat malware, or seek professional help. But if we want to understand how a computer virus or other malware will spread in a network of computers, we have to focus on the interaction between the malware and uncompromised computers and on the interactions between uncompromised and compromised computers.

## 1.3.   Our Long Term Goals, Pasteur's Quadrant, and the Resilience Stack

I have these long term goals in mind. First, to use key ideas from population biology to develop a systems dynamics model to explore the roles of attack and maintenance rates, detection capability, and other design parameters on the dynamics, particularly how compromise of the cyber system propagates, and performance of the enabled physical system responds. Second, with the goal of understanding what are the important variables and how they affect performance, our models will be **heuristic** ones that are not specific to any particular situation but have much in common with many cyber systems. Libicki *et al.* (2015) note that heuristic models allow us to increase understanding of important variables, rather to make precise predictions, and that such models provide a framework for thinking about cyber security choices (also see Mangel and McEver 2021). We seek to develop what Henderson and Taimina (2020) call active intuition, where "active" emphasizes that our intuition develops and grows by exploring the properties of our models and the predictions we obtain from them. Third, we will use the models to help identify what to measure to be able to assess vulnerability to cyber attack, the consequences of attack on performance of the cyber or enabled physical system, and to identify design tradeoffs and routes to defense.

These motivations are rooted in the particular application to cyber systems, but throughout we will seek fundamental understanding and this leads us to **Pasteur's Quadrant** (Stokes 1997, Mangel 2023).

### 1.3.1.   *Working in Pasteur's Quadrant*

In his book on technological change, *Pasteur's Quadrant*, Donald Stokes (1997) argued that a single axis between basic and applied science is the wrong way to think about the process of research. Rather, one must focus attention in a plane. One axis of assessment

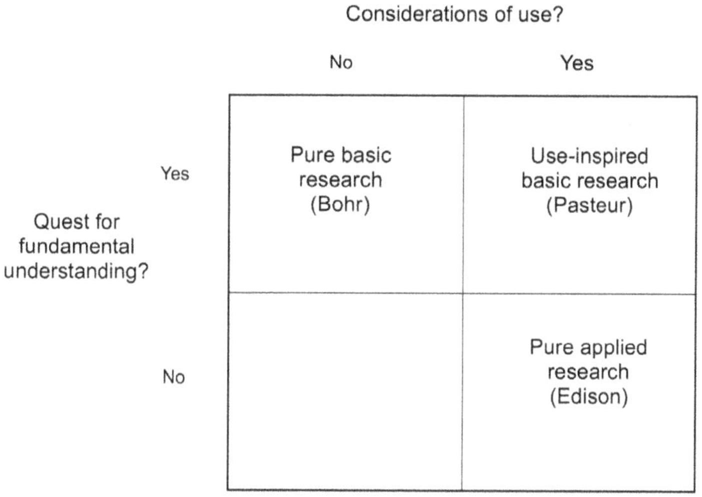

Considerations of use?

| | No | Yes |
|---|---|---|
| **Yes** | Pure basic research (Bohr) | Use-inspired basic research (Pasteur) |
| **No** | | Pure applied research (Edison) |

Quest for fundamental understanding?

**Fig. 1.3.** Stokes's (1997) vision of science is that in every scientific endeavor we may ask about application – is there consideration of use motivating this work – and whether or not there is a quest for fundamental understanding. Niels Bohr provides the canonical example of an individual whose work was not motivated by consideration of use but involved the deep search for fundamental understanding and Thomas Edison one whose work was motivated by consideration of use but did not search for fundamental understanding. Louis Pasteur's work from the time of his PhD was motivated by an important applied problem (Debré 1994) while he simultaneously sought fundamental understanding. Following Stokes, the left quadrant is unnamed; you might think what kind of work is done in that quadrant.

is whether the work is motivated by consideration of use and the other is whether there is a quest for fundamental understanding. The plane can then be divided into the four quadrants (Figure 1.3), (1) No consideration of use and quest for fundamental understanding; (2) Consideration of use and a quest for fundamental understanding; (3) Consideration of use and no quest for fundamental understanding; and (4) No consideration of use and no quest for fundamental understanding.

Stokes called the quadrant "not motivated by consideration of use and search for fundamental understanding" Bohr's Quadrant (although we are now clearly aware of the uses of Bohr's work understanding the atom), the one "motivated by consideration of use and no search for fundamental understanding" Edison's Quadrant, and the one "motivated by consideration of use and search for fundamental understanding" Pasteur's Quadrant. The fourth quadrant remains unnamed.

Pasteur's entire career involved work that was motivated by an important applied problem but simultaneously sought fundamental understanding (Debré 1994). Pasteur said "There is no such thing as a special category of science called applied science; there is science and there are its applications, which are related to one another as the fruit is related to the tree that has borne it". One great advantage of working in Pasteur's Quadrant is that we are most likely to develop transferrable methods when we seek fundamental understanding. Working in Pasteur's Quadrant is a theme of this book, since we are motivated by specific problems of cyber systems but seek fundamental understanding.

When considering a model, we should ask if the assumptions make sense and if they are consistent with what we know about the physical world. But, we should banish the word "realistic" because science by its very nature focuses on some things and ignores others, cannot capture all of reality. Caswell (1988), in a discussion of the roles of theory and models in ecology, gives a particularly interesting example. At a poster session on the terrestrial ecology of pine forests during the IVth International Congress of Ecology, he counted the number of posters in which 1, 2, 3, up to 6 factors were varied in the experiments. The vast majority of papers varied 1 or 2 factors, about which Caswell wrote "Surely no pine forest ecologist would argue that nutrient cycling is completely determined by two or three factors, and that all others are irrelevant ... the absence of other factors in the experimental design would be accepted as criticism of such a study only if it could be argued that those factors might change the answers to the questions is under investigation" (Caswell 1988, pp. 40–41).

So it is with modeling: we cannot model the entire world, so we must decide which factors are relevant to the questions that we ask (see Clark and Mangel (2000, Chapter 4) for a sequence of models in response to a sequence of more and more complicated questions). Then we need to ask about the fidelity of the assumptions to the natural world, whether the internal logic of the theory is such that the predictions flow from the assumptions or allows surprise predictions, and what happens when the predictions are confronted with data (Agutter and Tuszynski 2011, Serevido *et al.* 2014, Shou *et al.* 2015).

The **resilience stack** will help us frame the applied motivation, so we consider that next.

### 1.3.2.   *The resilience stack*

An attack on a cyber system first requires compromise by gaining access to cyber assets that interface with the external world. Once the attacker is inside the cyber system, the second stage of internal co-compromise can commence, in which compromised assets infect non-compromised assets. Compromised assets may immediately reduce the performance of the cyber system or the enabled physical system, or the adversary may hide compromise until ready to execute the attack. A successful cyber attack requires a weakness in software or hardware used by the defender, code that can exploit this weakness, and a method for propagating the exploiting code (Lachow and Grossman 2018).

In many cases, cyber assets that interface with external world can be protected by external hardness (Libicki *et al.* 2015) in which anti-malware prevents attackers from entering the cyber system (Gartzke and Lindsay 2015). Cyber assets that do not interface with the external world can similarly be protected from co-compromise by internal hardness. External and internal hardness rely on a variety of mechanisms (Gartzke and Lindsay 2015, National Academies of Sciences, Engineering, and Medicine 2017, Carlin 2018).

However, it is now clear that external and internal hardness are necessary but not sufficient for both a reliable and resilient cyber system (National Academies of Sciences, Engineering, and Medicine 2017). That is, we should assume that attackers will get through defenses (Libicki *et al.* 2015). In this case, resilience of the cyber system (and thus performance of the cyber system or the enabled physical system) requires some form of Defensive Counter Measure (DCM, *sensu* Mangel and McEver (2021)) that returns the system to a state closer to the one before the attack. Protection from external compromise and internal co-compromise may not be effective (e.g. the installed anti-malware does not defend) or may lose its effectiveness over time (e.g. the attacker discovers a way to circumvent the anti-malware currently installed).

Goldman (2020, p. 40) notes that "The core research question . . . is how to deter cyber attacks when you cannot reliably defend against them". The resilience stack allows a way to frame the answer to this question. Cyber systems are hierarchical. Clearly the best defense is to avoid attack in the first place. If one cannot avoid attack, then cyber assets can be dedicated to preventing the attack from entering the cyber system, i.e. resisting attack. When an attack cannot be resisted, i.e. the attacker enters the cyber system, one can prepare by hardening the cyber assets so that they are robust to attack, and by having redundant cyber assets, so that if one cyber asset fails others can take over the mission. But even hardened, redundant systems may fail against significantly sophisticated cyber tools so that assets will need to be repaired, reset, or even replaced. Replacement is a form of modularity by which the cyber system can be continue to be resilient to attack. Finally, when none of these methods work one needs to adapt by developing new assets or configurations of existing assets.

Farsangi *et al.* (2019), in a book on resilience in a wide range of physical systems, note that the concept of resilience has received considerable attention since 2010 because of the recognition that hazards and threats – both physical and cyber – cannot perpetually be averted. However, they also note that there is considerable debate on how to define resilience, which can have many meanings. Farsangi *et al.* (2019, p. v) choose to define resilience as "the capability of a system of a system to maintain or promptly recover its functionality in the face of extreme events".

In cyber systems, McEver *et al.* (2019) call the hierarchy described above the Resilience Stack (Figure 1.4), where resilience is the property of a system to function in the face of threats, rather than fail due to them (Singer and Friedman 2014). A resilient system, as characterized in Figure 1.4, will attempt to resist attack, but then will be able to function under degraded conditions caused by an attack, and will recover following the attack. Clearly, if the system can recovery quickly enough from a cyber attack, there is less concern than if the recovery is very slow (Singer and Friedman 2014).

We focus on resistance, recovery, and robustness and explore how process based models from population biology inform thinking about these issues. These parts of the resilience stack have a natural analogue in the population biology of disease. Organisms can avoid

## The Resilience Stack

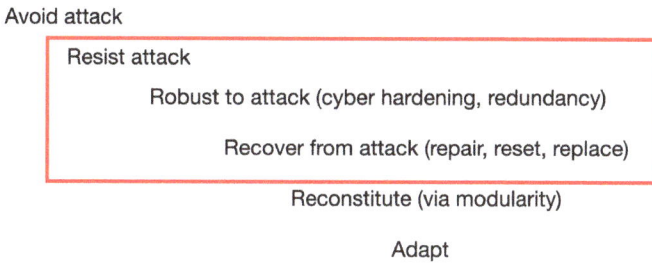

Avoid attack

Resist attack

Robust to attack (cyber hardening, redundancy)

Recover from attack (repair, reset, replace)

Reconstitute (via modularity)

Adapt

**Fig. 1.4.** The hierarchical nature of resilience in cyber systems (slightly modified from McEver *et al.* (2019)). Clearly it is best to avoid attack in the first place. If one cannot avoid attack, then the attack can be resisted. Even with resistance to attack, some attacks will succeed, so that one needs cyber systems that are robust to attack. Even with robustness, some attacks will succeed, so that one needs the ability to recover from the attack. It is also possible that the attack will be so severe that one needs the ability to reconstitute and ultimately adapt the system. The foci of our work are resistance, robustness, and recovery, highlighted in the red box.

attack (Ruxton *et al.* 2018) and if attacked can either resist (i.e. control) the disease organism or tolerate the damage caused by it (Read *et al.* 2008, Råberg *et al.* 2009, Merrill *et al.* 2021, Wilber *et al.* 2024). Similarly, Singer and Friedman (2014, pp. 155–156; pp. 174–177) argue that the we should strive to build cyber systems "where the parallel for measuring offense and defense isn't war, but biology... the body has built up a capacity of both resistance and resilience, fighting off what is most dangerous and, as Vint Cerf puts it, figuring out how to 'fight through the intrusion'... Just the mere existence of such a system [a defense that can outsmart the adversary] would always sow doubt in the offense that the attack is going to work". We should also expect that resistance, tolerance, and resilience depend upon the characteristics of the cyber system and the kinds of perturbations it experiences (for ecological examples, see Roopnarine *et al.* 2022, Sherrat and Stefan, 2024). We will see how quantitative models can make our intuition about resilience more precise.

Libicki (2016, p. 27ff, particularly his Figures 4.1 and 4.2) provides a metaphor for the resilience stack. A cyber system can be "noise-tolerant" (agoras) or "noise intolerant" (castles). About the former, he writes "With agoras, the risk from a single piece of bad information is low, but the benefit of having day-to-day access to the world of information is comparatively high (think Wall Street)... Recognizing that most individual pieces of information are of little use, possibly false, and usually transient, processes are defended from corruption by putting out more lines to the rest of the world. New information can be used to evaluate the old information, and the sophisticated agglomeration of information is the best path to making good decisions." About the latter he writes "With castles, the risk from bad information and hence bad instructions is high, but the benefit from having day-to-day access to the world of external information is comparatively low (think nuclear plants)". Like a physical castle, a cyber system that is a castle "protects itself through a series of enclosures: the open field of fire, the moat, the wall, and within the castle, the keep. Each obstacle challenges the intruder, and when one is breached, attention shifts to maintaining the next obstacle and, if possible, throwing the attacker back beyond the earlier breached obstacles".

In short, we are wise to design cyber systems with the assumption that they will be breached and the attacker with then have access to large parts of or all of the system. In general, as systems grow more complex there are more routes to failure, and the resilience stack thus grows in importance. Recognizing this fragility in complex systems encourages us to think about how to make them anti-fragile so that they can possibly gain from disorder (Taleb 2012), which is a topic beyond the scope of this book.

## 1.4.    The Dynamics of Cyber Attack

Often, once a cyber tool is used for an attack, the defender will be able to explore its properties and develop defenses that defeat that particular cyber tool, so that cyber tools are weapons whose effectiveness often rapidly decreases in time (Dipert 2010, Gartzke 2013, Smeets 2017). The attacker then needs to develop a new tool, which may take considerable time. In such a case, cyber attacks will come in pulses or waves. To incorporate these pulses in a mathematical model requires a mathematical description of the pulses. We model a single **pulse attack** using the classical Gaussian distribution (bell shaped curve), which requires specifying a mean $t_{peak}$ (which is the time at which the pulse peaks) and dispersion (standard deviation), which we denote by $\sigma$.

We adopt the symbol $I(t)$ to denote the **relative intensity of a pulse attack** as a function of time (put aside the SIRS or SIR models in which the same symbol is used for the number of infected individuals), writing it as

$$I(t) = \frac{1}{\sqrt{2\pi}\sigma} e^{-(t_{peak}-t)^2/2\sigma^2} \qquad (1.1)$$

Equation (1.1) characterizes the relative intensity of the attack, and we multiply it by the **attack rate parameter** $a$ to determine the absolute intensity of the attack, $aI(t)$. Setting $t = t_{peak}$ in Eqn. (1.1) that the maximum intensity of the attack is $a/\sqrt{2\pi}\sigma$.

The properties of the Gaussian distribution are well tabulated, so that we can readily assign times to the putative start and end of the attack. For example, 99.7% and 95.5% of the probability of the Gaussian distribution in Eqn. (1.1) is contained in $[t_{peak} - 3\sigma, t_{peak} + 3\sigma]$ and $[t_{peak} - 2\sigma, t_{peak} + 2\sigma]$ respectively. Thus, if we say the attack starts when its intensity is less than a fraction of percent of its the maximum intensity, the start time is $t_{peak} - 3\sigma$. Similar reasoning applies to the end time.

A persistent cyber attack is one in which the attacker always has a new tool ready when the current one fails. Persistent cyber operations between nations arise in part due to the cyber security dilemma (Buchanan 2016): as nations secure themselves they induce (not necessarily deliberately) fear in other nations. Even when the attacker does not always have a new tool ready, cyber deterrence is difficult, in part because of the difficulty in finding the correct response for a given cyber attack (Libicki 2018a), especially in public cyber institutions (Kostyuk 2021), and maximizing cyber security may require persistent operations in the cyber system of the adversary (Buchanan 2016, p. 52). There are also arguments for not relying solely on deterrence but for initiating persistent operations (e.g. Fischerkeller *et al.* 2020), including the possibility that if the attack rate is sufficiently low, it may be concealed from the adversary (Green and Long 2019). Thus, there is need to balance cyber offense and defense (Slayton 2016/2017).

For persistent operations, we will assume that intensity of the attack is much smaller than the maximum intensity of a single pulse attack. For example, we will sometimes set the persistent level of attack to be the average value of $aI(t)$.

Valeriano and Maness (2015, p. 82ff) constructed a Dyadic Cyber Incident Database (DCID) that allowed them to make a wide range of inferences about cyber attacks, which they broadly classify as either cyber incidents (111 in the DCID) or cyber disputes (45 in the DCID). Their cyber incident is more or less our pulse cyber attack and their cyber dispute is more or less our persistent cyber attack. They also note that although there 126 possible dyad interactions in the DCID, only about 20 rivals engage in cyber conflict.

We are thus led to conceive of the attack rate continuum (Figure 1.5 upper panel). In the next chapters, we will focus on the extremes (pulse attack, persistent attack) because they are amenable to analysis.

However, many situations will involve the intermediary situation in which attack consists of a series of pulses (Figure 1.5, lower panel). To model this situation, we consider a time horizon that runs from 0 to a maximum time $T$. During this time interval, we allow $J$ separate pulse cyber attacks. We then specify the peak $t_j$, standard deviation $\sigma_j$, and the maximum the intensity $a_j$ (which ranges from 0, i.e. the cyber attack is ineffective even at its peak, to a maximum value $a_{\max}$) of the $j$th pulse attack.

With this notation, we the relative intensity of the $j$th attack is

$$I_j(t) = \frac{1}{\sqrt{2\pi}\sigma_j} e^{-(t_j-t)^2/2\sigma_j^2} \tag{1.2}$$

and total intensity of attack is

$$I_T(t) = \sum_{j=1}^{J} a_j I_j(t) \tag{1.3}$$

In the lower panel of Figure 1.5, $a_j$ and $\sigma_j$ are random variables. Then each choice of the $a_j$ and $\sigma_j$ will generate a different realization of the pulse attack even if the peak times are the same. If you would like the pulse to be flatter at the peak, experiment by changing the power of $(t_j - t)$ from 2 to something larger, but be warned that

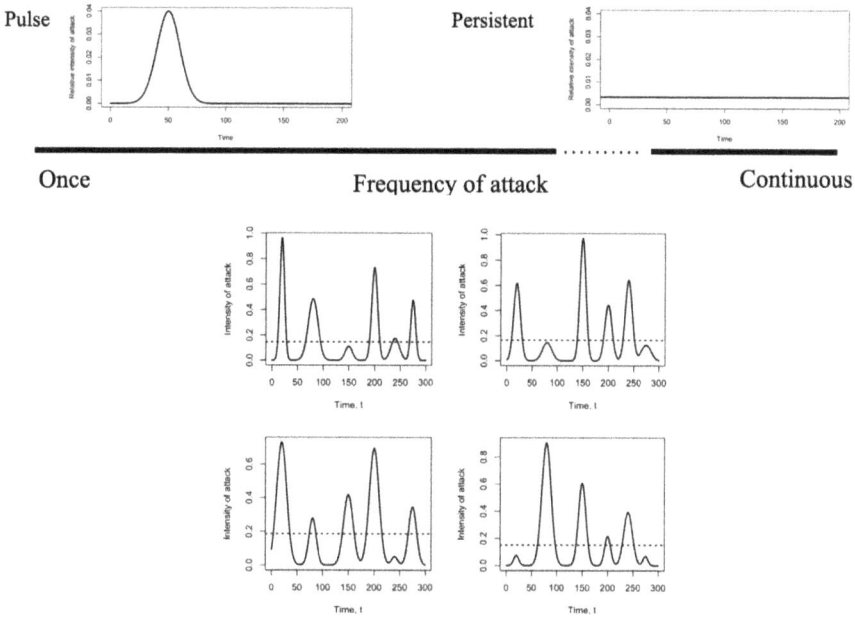

**Fig. 1.5.** At the extremes (upper panel), a cyber attack occurs only once, with a pulse characterized by Eq. (1.1) or continuously with attack rate given by the average value of the pulse. Most situations (lower panel) will sit between these cases and consist of a series of pulses. The extremes of either a pulse attack or continuous attack are amenable to analysis, while the intermediary cases are best studied by simulation; four different realizations of the multiple pulse attack are shown here.

the normalization constant $1/\sqrt{2\pi}\sigma_j$ also needs to change. When there are multiple pulses, as in the lower panel, it is reasonable to expect that the frequency and the intensity of the attack will interact as they affect the state of the cyber system (Miller *et al.*, 2011, 2012). This is a really interesting question, but also beyond the scope of this book and one that can be answered using the tools we develop.

> **Potential project**: Generate analogues of the lower panel in Figure 1.5 for pulse attacks that are wider than the Gaussian model and for which the parameters of the attack are correlated with each other.

### 1.4.1.  *Two examples of cyber attack: More about Stuxnet and Volt Typhoon*

Now that we have a vision of the pulse attack, I want to pause with the formulation of the model to discuss two examples of cyber attack.

#### 1.4.1.1.  *More about Stuxnet*

Zetter (2014) describes Stuxnet as the world's first digital weapon, aimed against centrifuges that Iran was using to enrich uranium, and gives a detailed account of its development, mechanisms, and consequences. Zetter gets into detailed mechanisms, such as zero-day exploits and reinfection of scrubbed computers by worms hiding in the network, that we will forgo.

The Stuxnet virus attacked a specific line of Siemens Programmable Logic Controllers (PLCs) of the Iranian centrifuges. PLCs are "small computers, generally the size of a toaster, that are used in factories around the world to control things like the robot arms and conveyor belts of assembly lines" (Zetter 2014, p. 17). For most of the book we do not explicitly model the cyber system of the enabled physical system, but return to it in Chapter 8 as a direction of future research.

Control systems of the enabled physical system are vulnerable in part because such systems "don't get replaced for years and don't get patched on a regular basis the way general computers do. The life-span of a standard desktop PC is three to five years, after which companies upgrade to new models. But the life-span of a control system can be two decades. And even when a system is replaced, new models have to communicate with legacy systems, so they often contain many of the same vulnerabilities as the old ones." (Zetter 2014, p. 147). There are many similar control systems (e.g for water-treatment and power plants, dams, bridges, and train stations, smart TVs and power meters, uninterruptible power systems) connected via the Internet of Things. For example, in 2012 a researcher in the UK identified more than 10,000 connected control systems (Zetter 2014, p. 148); since 2012 the number of connected control systems has grown enormously.

The first thing that Stuxnet did after getting into a machine was to explore the software and if "Stuxnet found itself on a system that didn't have the Siemens software installed, it simply shut itself down. It still sought other machines to infect, but it wouldn't launch its payload on any machine that didn't have the Siemens software installed. Any system without the software was just a means to Stuxnet's end." (Zetter 2014, p. 28). It is likely that Stuxnet was launched against a group of five companies in Iran, leading to 12,000 infections at those initial targets that spread to more than 100,000 machines in more than 100 countries (Zetter 2014, p. 97). Once Stuxnet infected a controller, the compromise took time to become apparent, in part because Stuxnet first conducted surveillance of the infected cyber asset, in part to convince the human operators that all was okay. Thus, there was a period of time (ranging from 35 to almost 300 days, depending upon the intensity of the attack) before damage to the centrifuges (increasing internal pressure) began in earnest.

However Stuxnet, which had taken years and considerable investment to develop, only lasted a few weeks before it was understood and rendered ineffective (of course, by then the damage to the centrifuges had been done) (Zetter 2014, p. 178, 318).

### 1.4.1.2.  *Volt Typhoon*

A May 2023 post by the Microsoft Threat Intelligence Team (Microsoft 2023) described the attack by the Volt Typhoon botnet. After investigation, Microsoft staff concluded that the Volt Typhoon attack was sponsored by China with the goal of disrupting critical communications infrastructure between the United States and Asia in the future and that the attack began in the summer of 2021. Volt Typhoon targeted critical infrastructure in Guam and the United States. The targets included communications, manufacturing, utility, transportation, construction, maritime, government, information technology, and education sectors. From the characteristics of the attack, Microsoft concluded that the attacker's goal was to perform espionage and maintain access without being detected for as long as possible. One of the motivations for the release of the May 2023 post was to bring the attack to the intention of the broader community.

Volt Typhoon is an example of stealth, hands-on keyboard activity and what Microsoft calls "living off the land" https://www.microsoft.com/en-us/security/blog/2018/09/27/out-of-sight-but-not-invisible-defeating-fileless-malware-with-behavior-monitoring-amsi-and-next-gen-av/). Volt Typhoon used the command line to (i) collect data, including credentials from local and network systems, (ii) archive those data for exfiltration, and (iii) persist in the target system using the stolen valid credentials. Additionally, according to Microsoft, Volt Typhoon "tries to blend into normal network activity by routing traffic through compromised small office and home office (SOHO) network equipment, including routers, firewalls, and VPN hardware" and used open-source tools to hide itself. Briefly, the attack proceeded as follows:

- Volt Typhoon achieved initial access through external nodes of the cyber network interfacing with the outside world and then tried to use privileges from a compromised system to extract the credentials of the currently attacked device and gain access to other components of the network with the stolen credentials.
- Once initial compromise was achieved, the Volt Typhoon human attackers conducted keyboard operations through the command line of the compromised software. Microsoft notes that "some of these commands appear to be exploratory or experimental, as the operators adjust and repeat them multiple times".
- That is, rather than employing malware in post-compromise operations, the attackers used "living-off-the-land commands" (as military forces do) to "obtain information on the system, discover additional devices on the network, and exfiltrate data".

The Microsoft report goes into more details – including examples of code – for access to credentials, discovering system information, collection of information from web browser applications, and how the Volt Typhoon command and control operated. The report then turns to specific mechanisms for defense against Volt Typhoon, again with specifics of code to illustrate how defense works. These specifics of attack and defense are beyond the scope of our work.

### 1.4.1.3.    *Are these pulse attacks?*

Are Stuxnet and Volt Typhoon pulse attacks? Clearly not, in the sense that Eqn. (1.1) is not a precise representation of the attacks. On the other hand, both attacks had a beginning, and a peak, and an end, as in Eqn. (1.1). Thus, we might say that model for the pulse attack in Eqn. (1.1) captures key features of Stuxnet and Volt Typhoon, while leaving many other features to be elaborated. This is part of the art of mathematical modeling – deciding which features of the world are key to communicate, and then finding a way to communicate ideas about those features. Thus one of your goals is to start developing the art of modeling, knowing that at the end of this book there will still be much to be done to resolve your discontent with Eqn. (1.1) (if you have any) and other models that we will develop.

## 1.5. Performance of the Cyber System or the Enabled Physical System

Cyber systems exist to do something, which may range from a collection of PCs networked at night for parallel computations (in my research group we did this many years ago), through communications systems, to enabling a physical system such as a power plant, aircraft guidance system, or even a car. In the case of the physical system, for a particular application one needs to model the specifics of that system. For example, Mangel and McEver (2021) used a coupled oscillator model of the electric grid to understand how compromise of smart meters can lead to grid failure via reverse engineering, but also introduced a more generic model of performance (also used in Mangel and Brown 2022). In this generic model, Mangel and McEver (2021) assumed that **performance** of the cyber system or the enabled physical system was determined by the numbers of uncompromised and compromised cyber assets.

In this book, we use a simpler metric of performance, assuming that performance depends only on the number of uncompromised cyber assets. In particular we assume that the relative performance of the cyber system or the enabled physical system in its usual tasks is measured on a scale from 0 to 1, and depends only on the number $x$ uncompromised cyber assets, modeled as

$$\phi(x) = \left[ \frac{1}{1 + e^{\frac{x_{50} - x}{\sigma_x}}} \right] \tag{1.4}$$

This performance function depends on the two parameters $x_{50}$ and $\sigma_x$. Since the exponential in the denominator is 0 when $x = x_{50}$, making the fraction equal to $1/2$, $x_{50}$ is the number of uncompromised assets giving performance of 50%.

The parameter $\sigma_x$ determines how rapidly performance rises as the number of uncompromised cyber assets increases. When $\sigma_x$ is small compared to the range of $X$ values, performance will rise sharply (be "knife-edged"), as with the black line in Figure 1.6. As $\sigma_x$ becomes larger, the knife-edge smooths to a S-shaped curve, which henceforth is called sigmoidal. We will call $\sigma_x$ the performance shape parameter.

We now have a clear quantitative description of the concept of redundancy in the Resilience Stack. Consider, for example, the black curve in Figure 1.6. Here the same performance is achieved as long as there are more than about 500 uncompromised cyber assets. For the green curve, we can draw the same conclusion as long as there about more than 600 uncompromised cyber assets, but for the blue curve there have to be more than 700 uncompromised assets for high performance. This is a major insight: we cannot talk about redundancy unless we have in mind how the cyber system is to be used and how to evaluate its performance.

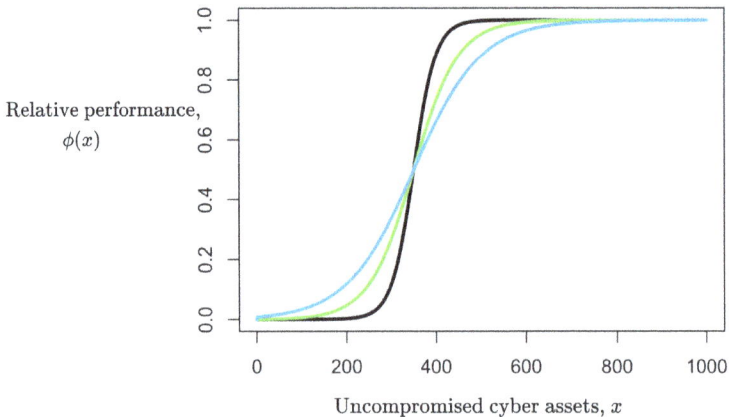

**Fig. 1.6.**  The performance function $\phi(x) = \left[ \dfrac{1}{1+e^{\frac{x_{50}-x}{\sigma_x}}} \right]$ depending upon the number of uncompromised cyber assets, $x$, $x_{50}$, at which performance is 50%, and the dispersion parameter $\sigma_x$ that characterizes the rise of performance from a small value to close to 1. For this figure, the parameters are $x_{50} = 350$ and $\sigma_x = 25, 50,$ or 75 (black, green, or blue lines, respectively. See the text for a discussion of how we can now capture the notion of redundancy from the Resilience Stack.

The shape of the three curves in Figure 1.6 depends upon the relative values of $x_{50}$ and $\sigma_x$ when compared to the total number of cyber assets. If $X_T$ denotes the total number of cyber assets, the shapes of the curves and the $y$-axis in Figure 1.6 will be the same as long as $x_{50} = 350 X_T/1000$ and $\sigma_x = 25 X_T/1000, 50 X_T/1000$, or $75 X_T/1000$ but the $x$-axis will now range from 0 to $X_T$. If this is not immediately clear, try explaining Eqn. (1.4) out-loud to yourself or to a colleague.

> **Potential project**: Eqn. (1.4) is only one of many ways to generate sigmoidal or S-shaped curves. Another of my favorites is $\phi(x) = \frac{x^\gamma}{x_{50}^\gamma + x^\gamma}$, with two parameters $x_{50}$ and $\gamma > 0$. Without numerical computation provide an interpretation for $x_{50}$. (If you are new to this kind of thinking, asks what happens when $x = x_{50}$.) Then explore the shape of $\phi(x)$ for different values of $\gamma$ (say $\gamma = 2, 4, 6, 8, 10$).

Clearly, when one has a particular application in mind, it is better to construct a specific model for the performance of the cyber or enabled physical system. In this book, we will mainly use Eqn. (1.4) or slight variations, but here is something to keep in mind as you read forward: cyber systems are often linked to one another and for a physical system to be enabled by a cyber system, the physical system will often have its own cyber assets. Such cases lead to linked performance functions.

For example, suppose that the performance of the enabled physical system when the performance of the cyber system is $\phi(x)$, denoted by $\mathcal{P}(\phi(x))$, is similar to Eqn. (1.4)

$$\mathcal{P}(\phi(x)) = \left[ \frac{1}{1 + e^{\frac{\phi_{50} - \phi(x)}{\sigma_\phi}}} \right] \tag{1.5}$$

where $\phi_{50}$ is the value of performance $\phi(x)$ of the cyber system giving 50% performance of the enabled physical system and $\sigma_\phi$ is a dispersal parameter capturing the rapidity of the rise in performance as $\phi(x)$ ranges from values near 0 to values near 1. In Figure 1.7, I show one visualization of the linkage between the focal cyber system and the cyber system of the enabled physical system. We will return to these ideas in Chapter 8.

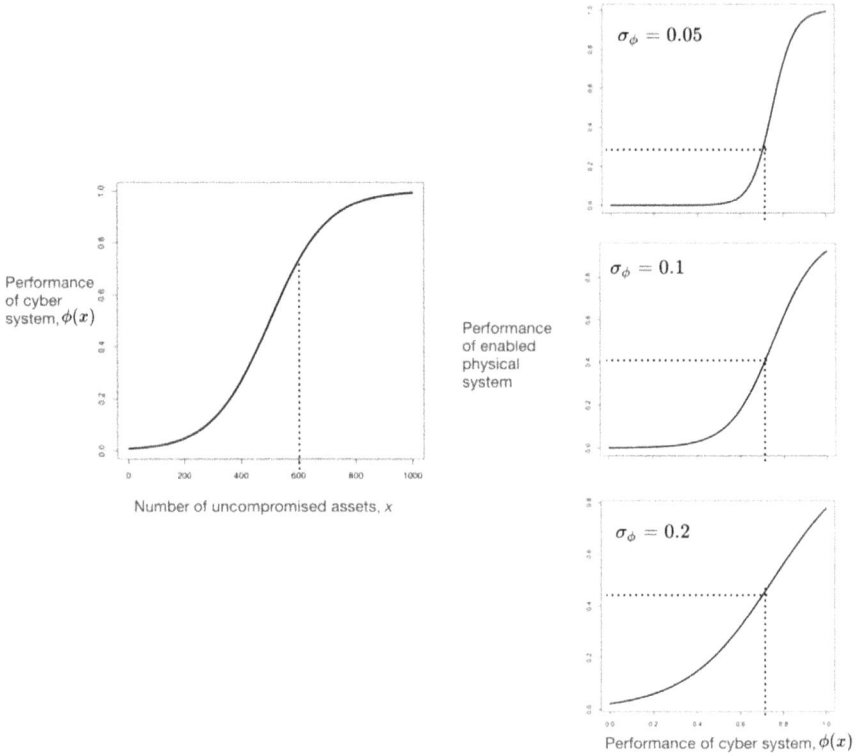

**Fig. 1.7.** The linkage between the performance of a cyber system and an enabled physical system that also has cyber assets. The left figure shows the performance of the cyber system $\phi(x)$ as a function of the number of uncompromised cyber assets. Note that $\phi(x)$ is on the $y$-axis. This links to the cyber assets of enabled physical system through performance $\mathcal{P}(\phi(x))$ through Eqn. (1.5), so that performance of the cyber system is now on the $x$-axis, as in the right figures. In those figures, $\phi_{50}$ is 0.75 (i.e. performance of the cyber system has to be at least 75% to obtain 50% performance of the enabled physical system). The three panels on the right show performance of the enabled physical system when the dispersal parameter is 0.05, 0.1, or 0.2. To illustrate the linkage, assume that 600 cyber assets as uncompromised. We then draw a vertical line to the curve in the left panel, showing that performance of the cyber system is slightly more than 0.70. In the three right panels, we then draw a vertical line at 0.7, let it intersect the curve and then draw a horizontal line to the $y$-axis in order to read off the performance of the enabled physical system.

## 1.6.    An Example of the Metaphor: Reverse Engineering of Advanced Metering Infrastructure to Compromise the Electric Grid

In Figure 1.8, I show the of the cyber system model that we used (Mangel and McEver 2021) to study how reverse engineering via compromise of Advanced Metering Infrastructure (AMI) can lead to instability in electric grids.

In this model of the cyber system, there are five kinds of cyber assets:

- Uncompromised and vulnerable cyber assets can be compromised either externally or internally. A fraction $\eta$ of these cyber assets are **decoys** with no functionality but instrumented to detect compromise with high probability.
- Uncompromised cyber assets that are currently invulnerable to either external or internal compromise are temporarily protected against malware, but as time progresses their anti-malware software ages and is no longer effective.

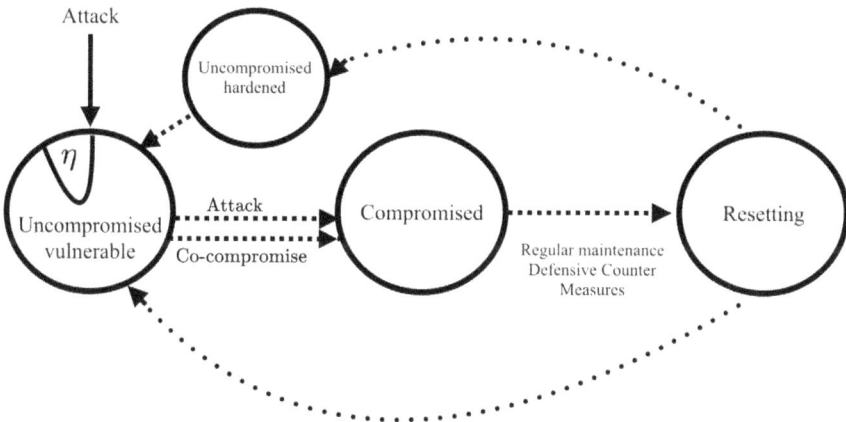

**Fig. 1.8.**   The cyber system model in Mangel and McEver (2021) contains cyber assets that are uncompromised and vulnerable, uncompromised and currently invulnerable (hardened), compromised, and resetting/restoring (and thus temporarily unavailable system; metaphorically "in the shop"). A fraction $\eta$ of the cyber assets (shown only for uncompromised and vulnerable assets) are decoys that do not contribute to functionality of the physical system and are unable to co-compromise.

- Compromised cyber assets are infected by malware.
- Once compromise is detected, defensive counter measures are activated to discover and send compromised assets to the resetting/restoring pool.
- Cyber assets in the resetting/restoring pool do not contribute to performance of the cyber system or the enabled physical system.
- After some amount of time, cyber assets that are being reset/restored return to the cyber system temporarily invulnerable (effective anti-malware installed) or still vulnerable (either no anti-malware installed or the installed anti-malware is ineffective).

Each of the cyber assets and the probability of detecting compromise has dynamics that must be modeled, leading to a series of ordinary differential equations. When these dynamics are coupled to a metric of performance of the cyber system or the enabled physical system, we are able to study the dynamics of compromise and how it affects performance. We (Mangel and McEver 2021) coupled the model of compromise of the cyber system to the synchronous motor model of an electric grid (Filatrella *et al.* 2008, Liu *et al.* 2013) to illustrate Electric Power Research Institute scenario AMI.27 (NESCOR 2013) for reverse engineering of Advanced Metering Infrastructure, showing how smart meters sending misleading signals about power demand can lead to load-side failure of the electric grid, and derive a condition for failure of the grid in terms of the number of compromised assets at the time of detection of compromise and the dissipation parameter of the synchronous motor model.

We begin (Chapters 2 and 3) with models that are simultaneously simplifications (with fewer pools of cyber assets) and extensions of the one shown in Figure 1.8. Then, we will slowly build additional complexity into the model of the cyber systems.

## 1.7.   Summary of Major Insights

- The Resilience Stack provides a way for us to think about the hierarchical nature of cyber attack and defense.
- We cannot talk about redundancy unless we have in mind how the cyber system is to be used and how to evaluate its performance or the performance of the enabled physical system.

- The methods of population biology are natural tools for understanding the dynamics of compromise and variability in cyber systems. Three key ideas from population biology relevant to cyber systems are:

  (1) Populations (of organisms or cyber assets) consist of individuals with different characteristics and successful modeling of the dynamics of populations must have level of description that matches the question of interest.

  (2) Populations have dynamics on many different time scales, but often reach steady or quasi-steady states (which may include periodic behavior [limit cycles]) in which dynamic processes are balanced.

  (3) Populations of organisms are governed by the fundamental law of biology – evolution by natural selection – acting on expected lifetime reproductive success of individuals as a proxy for the long-term representation of genes in the population. Although there is no similar fundamental law for cyber system, the notion of fitness maps into the performance of a cyber system or the physical system that it enables.

# Chapter 2

# The Pulse Attack Model

*A model should be as simple as it can be but no simpler*

– Albert Einstein

We now begin the modeling project in earnest, focusing on a single cyber attack. A pulse attack can have many origins. For example, the attacker may use a cyber tool for which the defender rapidly develops a defense, thus reducing the rate of attack. Alternatively, the attacker may simply want to demonstrate its capabilities or gather information about its adversary's network (Buchanan 2016, p. 130) and intentionally slow and then end the attack. Regardless, we will see that the long-term effects of a pulse attack depends on characteristics of the defender's cyber network.

In June 2022, after a visit to the Johns Hopkins University Applied Physics Laboratory, Charles Fraccia wrote to Christine Fox asking about epidemiological models of ransomware for the spread and infection of victim organizations. His thinking was that institution size and technology backbone would be "major factors that could be quantified (in such an epidemiological model) and help determine where to concentrate defenses at a specific time ... blunting the exponential dynamics of infectious models ... [and] would improve our defensive targeting capabilities rather that hoping for 0-full herd immunity model that we currently have" (quotation with permission of Charles Fraccia). Fraccia is the CEO of a company focusing on solving digital biosecurity challenges and improving manufacturing operations security, co-founder of an international organization that addresses threats unique to the bioeconomy and enables coordination among stakeholders to facilitate a robust and secure industry, and a member of the Information Science and Technology study group of DARPA (the US Defense Advanced Research Projects Agency). His question deserves serious attention.

The **Pulse Attack Model (PAM)**, whose basic version we develop in this chapter, is one answer to the questions raised by Charles Fraccia. As we go through the development, I expect that you will be thinking "but what about including . . .". Please be patient – in Chapter 5 we explore various extensions of the PAM. In this chapter, we assume that the defender knows the characteristics of the the pulse attack, and model the dynamics of cyber compromise, performance, and what the defender can do to mitigate the effect of the attack.

## 2.1.  Dynamics of Uncompromised and Compromised Assets

We assume that there is a single pulse attack, with relative intensity given by Eqn. (1.1) and total intensity $aI(t)$, where $a$ is the attack rate parameter, there are no decoy cyber assets, and after resetting all cyber assets return to the system unhardened and therefore vulnerable to attack once again (Figure 2.1).

We assume that the total number of cyber assets is constant, denoted by $X_T$. This in the analogue of the assumption in the population biology of disease that there are no deaths from the disease (Murray 2002, Merl *et al.* 2009). We relax this assumption in Chapter 5.

Cyber assets are then of three types

$$x(t) = \text{Number of uncompromised cyber assets at time } t$$
$$x_0(t) = \text{Number of compromised cyber assets at time } t$$
$$x_r(t) = \text{Number of resetting cyber assets at time } t \qquad (2.1)$$

Since the total number of cyber assets is constant, $x_r(t) = X_T - x(t) - x_0(t)$.

To determine the dynamics of these variables we assume

- Uncompromised assets are compromised at rate $aI(t)x(t)$.
- Once an asset is compromised, it can **co-compromise** other cyber assets and uncompromised cyber assets are co-compromised at rate $a_{co}x(t)x_0(t)$, where $a_{co}$ is the **co-compromise rate parameter**, which we delve into shortly.

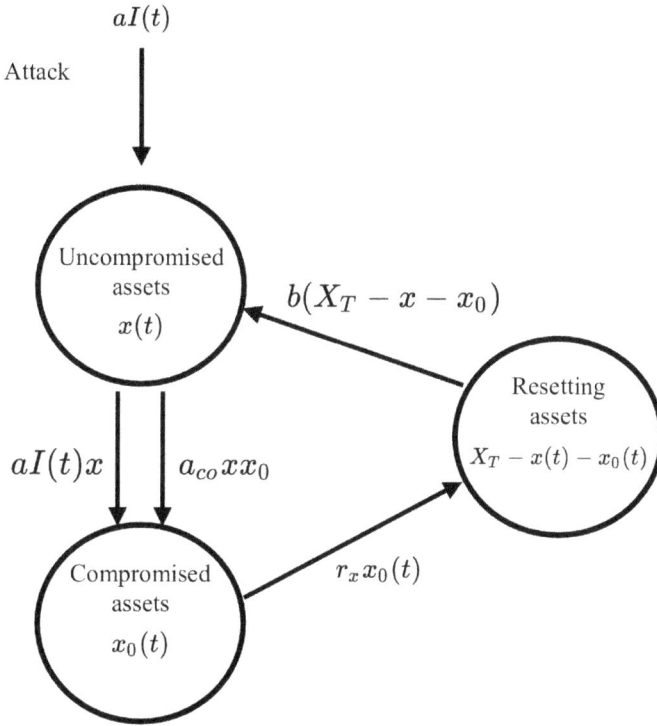

**Fig. 2.1.** In the basic Pulse Attack Model (PAM), we simplify the model in Figure 1.8 by eliminating the decoy cyber assets and assuming that all reset cyber assets return to the uncompromised pool vulnerable to attack. In contrast to Figure 1.8, the rate of attack is time dependent, with relative intensity given by Eqn. (1.1). We describe the cyber system by three pools of cyber assets (uncompromised, compromised, and resetting), which are symbolically denoted by the circles. Uncompromised cyber assets are attacked at rate $aI(t)$ and the arrow pointing from the attack to the uncompromised cyber assets shows that attack. Once compromised, cyber assets move to the compromised pool; this is shown by the arrow pointing downward. That compromised cyber assets can co-compromise uncompromised ones is captured by the second arrow pointing downward from the uncompromised pool to the compromised pool. The arrow pointing from the compromised pool to the resetting pool characterizes the removal of compromised cyber assets to resetting, and the arrow pointing from the resetting pool to the uncompromised pool shows the return of reset cyber assets to operational status. The number of cyber assets is dynamic, so that within the pools I have written $x(t)$ and $x_0(t)$.

- Compromised assets are moved to the resetting/recovery pool at rate $r_x x_0(t)$, either through specific detection of compromise or through regular maintenance.
- Cyber assets in the resetting pool are returned to the uncompromised state at rate $bx_r = b(X_T - x(t) - x_0(t))$, where $b$ is the rate or resetting/restoration.

Thus dynamics of the uncompromised and compromised cyber assets are

$$\frac{dx}{dt} = -axI(t) - a_{co}xx_0 + b(X_T - x - x_0) \tag{2.2}$$

$$\frac{dx_0}{dt} = axI(t) + a_{co}xx_0 - r_x x_0 \tag{2.3}$$

To further interpret the terms on the right sides of Eqns. (2.2) and (2.3) note that the units of the left sides of these equations are cyber assets per time, that is $[dx/dt, dx_0/dt] \sim$ *cyber assets/time*.

We already discussed $aI(t)$ in Chapter 1 (Eqn. (1.1)): $a/\sqrt{(2\pi)}\sigma$ is the maximum rate of attack occurring at $t = t_{peak}$ and $I(t)$ is the time dependent intensity of attack given by Eqn. (1.1). Uncompromised cyber assets are compromised at a rate proportional to the intensity of attack times the number of uncompromised cyber assets at the current time. Since compromised cyber assets appear at the same rate at which uncompromised cyber assets disappear, we have accounted for the first terms on the right sides of Eqns. (2.2) and (2.3). Since $I(t)$ is dimensionless and $xI(t)$ has units of *cyber assets* so that for $axI(t)$ to have units of *cyber assets/time* (in order to match the left of Eqn. (2.2)), $a$ must have units of $1/time$, making it a pure rate.

The middle terms on those right sides account for the process by which already compromised cyber assets $x_0(t)$ compromise the currently uncompromised cyber assets $x(t)$. Thus, $a_{co}$ characterizes the rate at which co-compromise of cyber assets occurs. But note that it is not a pure rate, because to match the units of *cyber assets/time* the units of $a_{co}$ must be $1/(cyber\ assets \cdot time)$. With these units the product $a_{co}xx_0$ has units

*[1/(cyber assets $\cdot$ time)] $\cdot$ cyber assets$^2$=cyber assets/time.*

When we envision uncompromised and compromised assets as components of a population, using their product can be called a "mass action" model, which has its origin in chemical kinetics (Coveney and Highfield 2023, pp. 118–119) and has been applied in population biology for more than century (e.g. Lotka 1924/1956, Gause 1934/2019, Kostitzin 1939, Goodstein 2007) with enormous success. During World War I, Frederick Lanchester pioneered the use of mass-action models in operational situation (Lanchester 1917) and they continue to be used to model the kinetics of combat (e.g. Taylor 1983, Keane 2011) and search (e.g. submarines looking for convoys or anti-submarine warfare aircraft looking for submarines (Wadington 1973,[1] McCue 2022)).

The third terms on the right sides of Eqns. (2.2) and (2.3) involve $b$ and $r_x$, which are also pure rates, with units $1/time$ characterizing how rapidly cyber assets that are being reset return to the uncompromised state ($b$) and how rapidly cyber assets that are compromised are removed from the compromised pool to be reset ($r_x$).

If all cyber assets are uncompromised at $t = 0$, the initial conditions for Eqns. (2.2) and (2.3) are $x(0) = X_T$ and $x_0(0) = 0$. These equations are readily solved by numerical routines for ordinary differential equations that are available in R, such as deSolve (Soetaert *et al.* 2018) and other languages have similar packages. For the computations in this chapter, I used the $4^{th}$ order Runge Kutta method in deSolve.

In Table 2.1, I summarize the symbols used in this model; their numerical values for computation are given in the caption to the first figure with numerical results.

Our general focus is about how ideas from the population biology of disease can inform cyber security, but sometimes we will look in the opposite direction. For example, recall that in our discussion of the SIR and SIRS models in the previous chapter, we noted that at least one infected individual has to be "injected" into a population of only susceptible and recovered individuals if there are going to be any interesting dynamics. That is, in analogy to Figure 2.1, the SIR and SIRS models need a pulse of infection to start the dynamics. This is usually done by starting with a positive number of infected individuals, but also could be caused by a pulse of disease agent,

---

[1]For readers who are biologists: this is the same Waddington of the epigenetic landscape.

**Table 2.1.**   Variables, Parameters and their Interpretation

| Symbol | Interpretation |
|---|---|
| $x(t)$ | Uncompromised cyber assets at time $t$ |
| $x_0(t)$ | Compromised cyber assets at time $t$ |
| $X_T$ | Total number of cyber assets |
| $a$ | Attack rate parameter |
| $a_{co}$ | Co-compromise rate parameter |
| $I(t)$ | Relative intensity of the cyber attack |
| $t_{peak}$ | Time at which relative intensity peaks |
| $\sigma$ | Standard deviation of relative intensity |
| $b$ | Rate at which cyber assets are reset and returned to operation |
| $r_x$ | Rate at which compromised cyber assets are moved to resetting |

as in the anthrax attacks in the United States about 25 years ago (Ingelsby *et al.* 2002, Wein *et al.* 2003), where the model of a pulse attacks fits nicely.

### 2.1.1.   *Interpretation of the co-compromise rate parameter*

We model the rate of co-compromise as proportional to the product of the number of uncompromised assets $x$ and the number of compromised assets $x_0$, with proportionality constant $a_{co}$. The co-compromise rate parameter $a_{co}$ is determined by the structure of the cyber network. Determining that rate for a network of a given structure is both important and amenable to modeling and empirical study. Doing so is beyond the scope of this book, but a good starting point is Easley and Kleinberg (2010). Clearly, the only way to ensure no co-compromise is that there are no links between the different cyber assets, which also means no passing of thumb drives between the operators of those assets. In this case, if one of the cyber assets in this network is compromised, it cannot pass the compromise to any other cyber asset. However, this is really not a cyber network, but a collection of independent cyber assets. Singer and Friedman (2014) describe this as "under-entitlement" of access and clearly $a_{co} = 0$. On the other hand, if every cyber asset is connected directly or indirectly to every other cyber asset, when one of them becomes compromised, unless action is taken, compromise will pass to all other cyber assets, regardless of which cyber asset is initially compromised. In this case $a_{co}$ will take is maximum value. Singer and Friedman (2014) describe this as "over-entitlement" of access.

There will be many intermediate cases. For example, the full network might be divided into subnetworks in which cyber assets are connected to each other but the subnetworks are not connected. In this case we expect $a_{co}$ to take an intermediate value between 0 and its maximum possible value. When the subnetworks have a connection between them, we expect that $a_{co}$ will be larger than it is when the subnetworks are not connected but smaller than the maximum value of $a_{co}$.

The value of $a_{co}$ for every network will sit between under-entitlement and over-entitlement of access, and determining the appropriate network structure is an important and non-trivial matter (but also beyond the scope of this book). When designing a communications network, for example, we might consider both the number of messages transmitted and their importance, the quality of the messages transmitted (e.g. what is the probability that a message is garbled), the cost of sending messages, and the probability that an important message is not transmitted.

> **Potential project**: Sketch out how you would draw the situations described in the above two paragraphs. Then think about how to explore how properties of the network such as the number of connections between one cyber asset and others and the time between compromise of an cyber asset and discovery determine the rate at which compromise is passed between assets. Can you design an experiment to measure properties of your networks?

Chapter 3 in Libicki (2016), on how to compromise a computer, provides many examples of the mechanisms of compromise and co-comprise and discusses the Apple individual Operating Systems (iOSs) for iPhones and iPads that provide a high level of defense and how it is achieved.

## 2.1.2. The differential equation for the relative intensity $I(t)$ of the pulse attack

When using deSolve In R, we need to append a differential equation for the pulse to Eqns. (2.2) and (2.3).

The equation of the pulse is

$$I(t) = \frac{1}{\sqrt{2\pi}\sigma}e^{-(t_{peak}-t)^2/2\sigma^2} \tag{2.4}$$

We differentiate Eqn. (2.4) using the chain rule from calculus:

$$\frac{dI}{dt} = \frac{1}{\sqrt{2\pi}\sigma}\frac{d}{dt}\left[e^{-(t_{peak}-t)^2/2\sigma^2}\right]$$

$$= \frac{1}{\sqrt{2\pi}\sigma}e^{-(t_{peak}-t)^2/2\sigma^2}\left[\frac{t_{peak}-t}{\sigma^2}\right]$$

$$= \left[\frac{t_{peak}-t}{\sigma^2}\right]I \tag{2.5}$$

We then append the equation $\frac{dI}{dt} = \left[\frac{(t_{peak}-t)}{\sigma^2}\right]I$ to Eqns. (2.2) and (2.3), with the initial condition determined by setting $t = 0$ in Eqn. (2.4), so that $I(0) = \frac{1}{\sqrt{2\pi}\sigma}e^{-t_{peak}^2/2\sigma^2}$.

## 2.2.  The Persistence of Compromise

Intuition suggests that once the attack starts the number of uncompromised assets will decline and the number of compromised assets will rise; numerical solution of Eqns. (2.2), (2.3), and (2.5) confirms that this is the case (Figure 2.2), with similar qualitative properties but quantitative differences according to the rate of co-compromise $a_{co}$.

But note that what happens *after* the attack ends is both qualitatively and quantitatively different. In the upper panel in Figure 2.2, after attack ends ultimately all compromise is removed, whereas in the lower panel compromise persists long after the attack has ended. This is something that we should understand and explain.

Long after attack has ended (so that the relative intensity of the pulse $I(t)$ is vanishingly small), the steady states of Eqns. (2.2) and (2.3) are obtained by setting $\frac{dx}{dt} = \frac{dx_0}{dt} = 0$. Using overline to denote the steady state, a little bit of algebra leads to

$$b(X_T - \overline{x} - \overline{x}_0) = a_{co}\overline{x} \cdot \overline{x}_0 \tag{2.6}$$

$$a_{co}\overline{x} \cdot \overline{x}_0 = r_x\overline{x}_0 \tag{2.7}$$

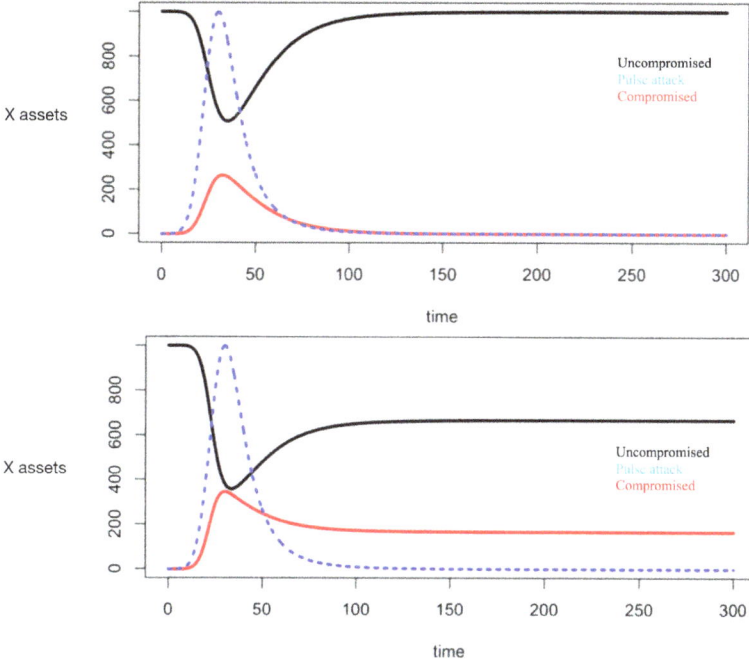

**Fig. 2.2.** The dynamics of uncompromised (black) and compromised (red) cyber assets when the relative intensity of attack is shown by the dotted blue line (scaled to $X_T \cdot I(t)/\max(I(t)$ to have comparable ordinate values). Common parameters to both panels are $a = 0.2, X_T = 1000, t_{peak} = 50, \sigma = 10, b = 0.2, r_x = 0.2$ The rate of co-compromise is $a_{co} = 0.0001$ in panel a) or $a_{co} = 0.0003$ in panel b. The major qualitative difference is the extinction or persistence of compromise following the end of the pulse attack. This is something we should explain.

Clearly, $\overline{x}_0 = 0$ is a solution to Eqn. (2.7) and when this is true Eqn. (2.6) implies that $\overline{x} = X_T$. This corresponds to Figure 2.2(a).

If $\overline{x}_0 \neq 0$, then Eqn. (2.7) implies that the steady state level of uncompromised cyber assets is $\overline{x} = r_x/a_{co}$. We use this in Eqn. (2.6) and solve for the level of compromised cyber assets to obtain

$$\overline{x}_0 = \frac{b(X_T - \overline{x})}{a_{co}\overline{x} + b} = \frac{b(X_T - \overline{x})}{r_x + b} \tag{2.8}$$

We interpret the right side of Eqn. (2.8) as follows: $X_T - \overline{x}$ is the number of cyber assets that are either compromised or being reset. The fraction of these cyber assets that are compromised is determined by the ratio $\frac{b}{r_x+b}$. When $r_x \ll b$ is very small (so that compromised cyber assets are moved into resetting very slowly), $\frac{b}{r_x+b} \approx 1$ and most of the cyber assets that are compromised or being reset will be in the compromised state. On the other hand, when $b \ll r_x$ is very small, $\frac{b}{r_x+b} \approx b/r_x$ and the rate limiting step in returning compromised assets to operational status will be resetting. In this case most of the cyber assets that are compromised or being reset will be stuck in the resetting pool, waiting to return to operational status.

The numerator on the right side of Eqn. (2.8) is positive only when $X_T > \bar{x}$ and since $\bar{x} = r_x/a_{co}$, we conclude that if $X_T > \bar{x} = r_x/a_{co}$ compromise will persist in a steady state long after the attack has ended.

That is, there is a threshold level of the co-compromise rate parameter

$$a_{co_{th}} = r_x/X_T \tag{2.9}$$

with the interpretation that if $a_{co}$ is less than this threshold, compromise will not persist in the steady state and if $a_{co}$ is greater than this threshold, compromise will persist in the steady state even absent external attack. For the parameters used to generate Figure 2.2, the threshold level of co-compromise is $a_{co_{th}} = 0.0002$, which is larger than the value of $a_{co}$ used in Figure 2.2(a) and smaller than the value of $a_{co}$ used in Figure 2.2(b).

This conclusion can be summarized in a plot showing the steady state level of uncompromised cyber assets as a function of the co-compromise rate parameter. When $a_{co} < a_{co_{th}}$, $\bar{x} = X_T$, and otherwise $\bar{x}$ smoothly declines as the rate of co-compromise increases (Figure 2.3, upper panel).

To explore the consequences of the co-compromise rate parameter on performance, we use the performance function in Eqn. (1.4) with $x_{50} = 500, 600$ or $700$ and once again sweep over a range of values for $a_{co}$ (Figure 2.3, lower panel). From the analysis in Eqns. (2.6–2.8) we know that when $a_{co}$ is less than the threshold value $r_x/X_T$ compromise will be 0 in the steady state following recovery from the attack. Hence for those values of $a_{co}$ we have $\bar{x} = X_T$. When the co-compromise rate parameter exceeds the threshold value, we have $\bar{x} = r_x/a_{co}$ (upper panel, Figure 2.3). Using these steady state values in the performance function $\phi(x)$ allows us to plot steady state performance as a function of the co-compromise rate parameter (lower

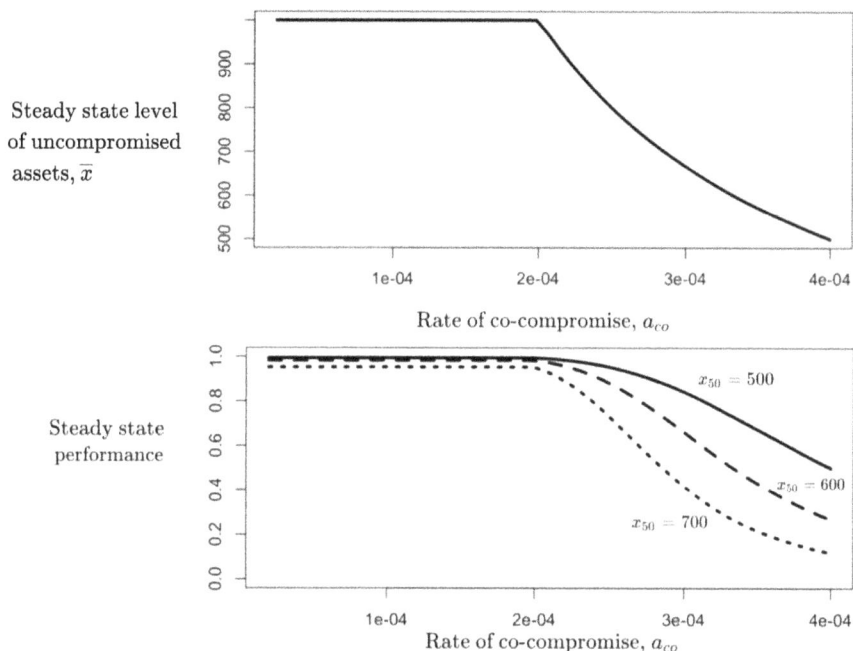

**Fig. 2.3.** Upper panel: Sweeping over values of the co-compromise rate parameter $a_{co}$ allows us to plot the steady state level of uncompromised cyber assets as a function of $a_{co}$. Lower panel: By combining the steady state value of uncompromised cyber assets with the performance metric in Eqn. (1.4) we are able to plot steady state performance as a function of the co-compromise rate parameter. In this panel $\sigma = 100$ and three values of $x_{50}$ are used in the performance function.

panel, Figure 2.3). Furthermore, note that even when compromise is extinguished long after the attack ends (i.e. $a_{co}$ is less than $2 \cdot 10^{-4}$), if $x_{50}$ is sufficiently large performance is close to but not exactly 1 (left side of the lower panel).

We thus conclude that in anticipation of cyber attack, the very first pro-active measure is to ensure that the co-compromise rate parameter is less than $r_x/X_T$. This is a property of the design of the cyber network, which is under the control of the defender. One can envision cyber attack "drills" by seeing how quickly "malware" moves through a network. Singer and Friedman (2014, pp. 211–214) explicitly call for such drills.

As a simultaneous check on our analysis and an interesting aside, we can compute the number of cyber assets in the resetting pool at the steady state when $a_{co}$ exceeds the threshold value (there are none when $a_{co}$ is less than the threshold because compromise does not persist). Since $\bar{x}_r = \bar{x}_r = X_T - \bar{x} - \bar{x}_0$. Using the far right side of Eqn. (2.8), a little bit of algebra shows that

$$\bar{x}_r = \frac{r_x}{r_x + b}(X_T - \bar{x}) \tag{2.10}$$

As a check that you both understand the analysis and can explain it, verbalize (preferably to a colleague), your interpretation of Eqn. (2.10) when $r_x \gg b$ or when $b \gg r_x$.

### 2.2.1.   *The analogue in the population biology of disease\**

Readers who are familiar with the population dynamics of disease know about the concept of a Host Threshold Density (HTD) that is required for disease to persist in a population. In this section, we will briefly explore the connection between the HTD and the threshold level of co-compromise.

We consider the Susceptible($S(t)$)-Infected($I(t)$)-Recovered ($R(t)$)-Susceptible (SIRS) model with a closed population (such as a cold or non-fatal flu going through a boarding school (Murray 2002)) so that $S(t) + I(t) + R(t) = N$, the total population size. Here are the SIRS equations once more

$$\frac{dS}{dt} = -\beta IS + \gamma R$$

$$\frac{dI}{dt} = \beta IS - \mu I$$

$$\frac{dR}{dt} = \mu I - \gamma R$$

where $\beta$ characterizes the rate at which already infected individuals infect susceptible individuals, $\gamma$ is the rate at which recovered individuals become susceptible to the disease again, and $\mu$ is the rate at which infected individuals move into the recovered pool.

Because of the assumption of the closed population, we can rewrite the dynamics of susceptible individuals as

$$\frac{dS}{dt} = -\beta IS + \gamma(N - I - S)$$

The steady states, again denoted by overline, satisfy $\beta \bar{I} \cdot \bar{S} = \gamma(N - \bar{I} - \bar{S})$ and $\beta \bar{I} \cdot \bar{S} = \mu \bar{I}$. One solution of these equations is $\bar{I} = 0$ and $\bar{S} = N$, in which case the disease is extirpated.

When $\bar{I} > 0$, so that the disease persists in the population, since $\beta \bar{I} \cdot \bar{S} = \mu \bar{I}$ we conclude $\bar{S} = \mu/\beta$ so that we can write the $\beta \bar{I} \cdot \bar{S} = \gamma(N - \bar{I} - \bar{S})$ as

$$\mu \bar{I} = \gamma \left( N - \bar{I} - \frac{\mu}{\beta} \right)$$

We solve this equation for the number of infected individuals in the steady state and find

$$\bar{I} = \frac{\gamma}{\mu + \gamma} \left( N - \frac{\mu}{\beta} \right) \tag{2.11}$$

and conclude that the disease will persist if $N > \frac{\mu}{\beta}$.

This is the biological analogue of Eqn. (2.9). The mathematics leading to these equations is the same, but what we do with the mathematics differs because of the difference in the question about cyber and biological systems.

## 2.3. Design Tradeoffs Between The Discovery of Compromise and Restoration from Compromise

In addition to $a_{co}$, the parameters $b$ and $r_x$ in Eqns. (2.2) and (2.3) are ostensibly under the control of the defender. Furthermore both are pure rates; let us start thinking of them as **resources** with values chosen by the defender. In this section, we assume their values are constrained by an overall resource level. The most general assumption, if constraint is linear, is that $c_b b + c_r r_x = \mathcal{R}$, where $c_b$ and $c_r$ are costs of a unit of $b$ and $r_x$ respectively, and $\mathcal{R}$ is the total resource level. It is easy to envision more complicated constraints. For example, password changes and software updates are (should be) done on a regular schedule (rather than in response to infection) so that their cost is proportional to the total number of cyber assets. For ease of computation, in the sense that we do not have to introduce additional parameter values, I will set $c_b = c_r = 1$.

To explore the dynamics of the cyber assets and performance as $b$ and $r_x$ covary, I set $\mathcal{R} = 0.4$, which is the sum of $b$ and $r_x$

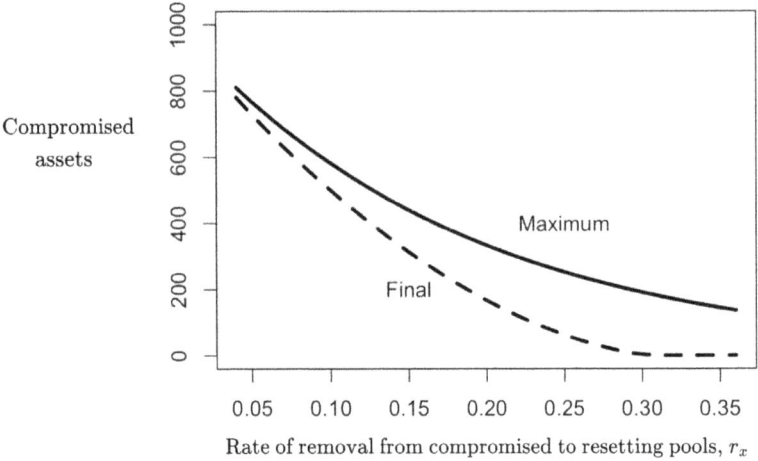

**Fig. 2.4.** A sweep over $r_x$, with $b = 0.4 - r_x$ allows us to track the maximum number of compromised cyber assets (solid line) and the final or quasi-steady state number of compromised assets (dotted line) as a function of $r_x$. In light of Eqn. (2.9) we expect that compromise will persist once $r_x$ falls below the value corresponding to the threshold level of co-compromise. We see this clearly in the dotted curve.

used to generate the results shown in Figure 2.2. I then used all the previous parameters, particularly the value of $a_{co}$ and let $r_x$ range from 0.04 (i.e. 10% of $\mathcal{R}$) to 0.36 (90% of $\mathcal{R}$), set $b = 0.4 - r_x$, solved Eqns. (2.2), (2.3), and (2.5) and used Eqn. (1.4) to assess performance with $x_{50} = 300$ and $\sigma_x = 200$.

First consider the dynamics of compromised cyber assets (Figure 2.4). Based on our understanding of the threshold level of co-compromise, we expect that as $r_x$ decreases, compromised cyber assets will persist in the steady state, long after the attack itself has ended. This is indeed the case (dotted curve in Figure 2.4). Furthermore, during the attack the maximum number of compromised cyber assets increases as $r_x$ decreases (upper panel in Figure 2.4).

The dynamics of uncompromised cyber assets (Figure 2.5) tell the complementary story. That is (lower panel), for sufficiently high $r_x$, there are only uncompromised cyber assets in the steady state and the steady state number of uncompromised cyber assets declines as $r_x$ declines. On the other hand, the minimum number of uncompromised cyber assets (upper panel in Figure 2.5) rises and then falls as $r_x$ increases. Thus $r_x$ and $b$ interact to determine the minimum number

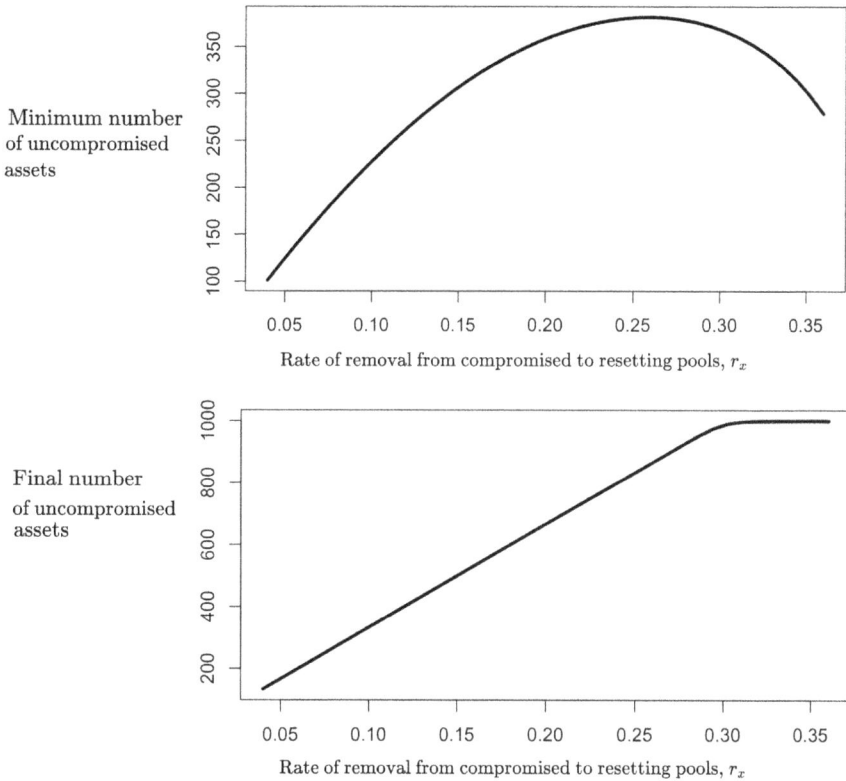

**Fig. 2.5.** The dynamics of uncompromised assets tell a story complementary to the one in the previous figure, with a twist. The complementary part is shown in the lower panel, in which the steady state (final) number of uncompromised cyber assets is $X_T$ is large enough, and otherwise declines as $r_x$ declines. The twist is shown in the upper panel, in which the minimum number of uncompromised cyber assets has a peak as $r_x$ varies from 10% of $\mathcal{R}$ to 90% of $\mathcal{R}$. This is our first hint of a design tradeoff in the choices of $r_x$ and $b$.

of uncompromised cyber assets. For example, for very small values of $b$ reset uncompromised cyber assets will accumulate in the resetting pool but only slowly return to the uncompromised pool.

Since performance is determined only by the number of uncompromised cyber assets, we expect that performance will mirror the number of uncompromised cyber assets. This is indeed the case (Figure 2.6). Indeed the upper panels of Figures 2.5 and 2.6 are virtually identical. Because of the sigmoidal nature of the performance function the linear relationship between the steady state level of

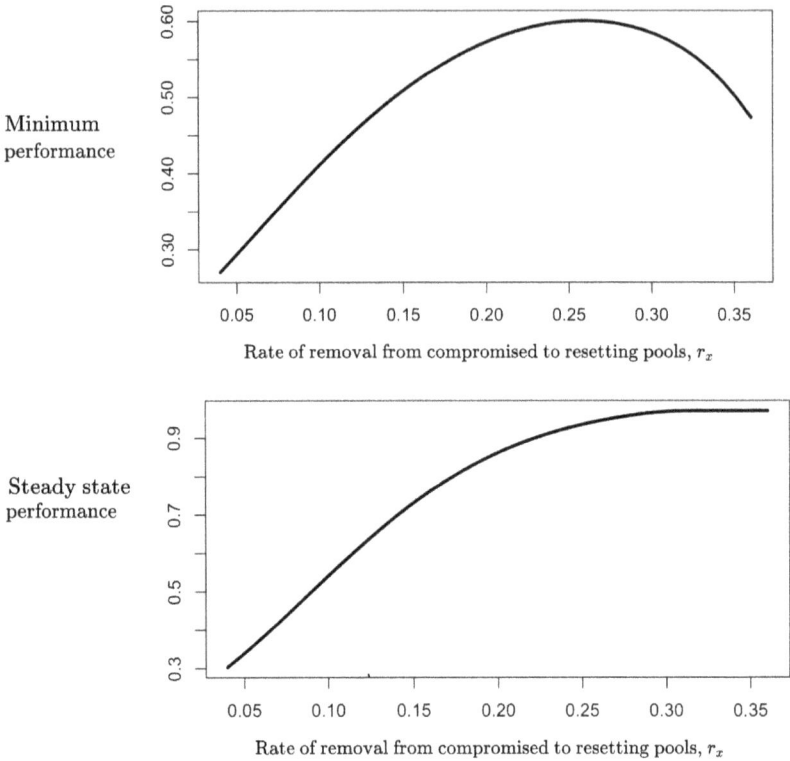

**Fig. 2.6.** Because performance depends only on the number of uncompromised cyber assets, both minimum performance (upper panel) and steady state performance (lower panel) are similar to the upper and lower panels of the previous figure.

uncompromised cyber assets and steady state performance becomes a nonlinear one. Note, for example, that when $r_x = 0.2$, the steady state number of uncompromised cyber assets is about 600, i.e. about 60% of the total number of assets but that performance is about 0.8, i.e. 80% of maximum possible performance.

These observations give us an inkling of the design tradeoff, assessing how $b$ and $r_x$ affect performance. Indeed, one can immediately think of all sorts of interesting maxi-min questions such as how to choose $r_x$, given the constraint between it and $b$. Perhaps even more interesting is the determination of a range of values of $r_x$ in which performance is pretty good. For example, if we were satisfied with performance greater than 0.9, examination of the lower panel in Figure 2.6

suggests that values of $r_x$ bigger than about 0.25 will do the job. We will return to this question in Chapter 5, when we consider extensions of the PAM.

Modeling the tradeoff between $r_x$ and $b$ allows us to understand how the discovery of compromise and the return cyber assets to the uncompromised pool interact. Libicki (2016, p. 84) considers that the most difficult question is the discovery of compromise. Intuition tells us that putting all resources into $r_x$ is likely to be unwise, and our results confirm this.

> **Potential project:** When $a_{co}$ is less than the threshold value, we know that compromise will ultimately be eliminated, so that we can define resilience following an attack as the time it takes for the cyber system to return to a high fraction of uncompromised assets. Develop a code to explore how such recovery time depends upon $r_x$, using the same constraint between it and $b$ that we have used. (We return to this question in Chapter 5.)

## 2.4. Summary of Major Insights

- Following a pulse attack, compromise may persist even when the system reaches a quasi-steady state long after the attack has ended. Whether this happens or not is determined by co-compromise rate parameter in the cyber system and this is a property of the cyber network under the control of the defender.
- Consequently in anticipation of cyber attack, the very first proactive defensive measure is to ensure that the co-compromise rate parameter is less than the threshold for persistence of co-compromise, which is a design parameter of the system.
- Cyber security drills can ensure that the co-compromise rate is below the threshold value by measuring the co-compromise rate parameter.
- There may also be design tradeoffs between the rate at which compromised cyber assets are discovered and removed to be reset and the rate at which they are returned to the uncompromised state. When this tradeoff exists, there is an optimal rate of removal to

resetting that minimizes the maximum level of compromise (thus maximizing the minimum level of performance) but also a range of values of rate of removal to resetting that is consistent with making performance pretty good during the attack while eliminating compromise after the attack ends.

# Chapter 3

# The Fundamental Model of Simultaneous Cyber Operations

*Cyber conflict is relatively new and the dynamics still relatively unknown. Any act of cyber deterrence is best thought of as an experiment. Some will work, some will not. The best hope is to think, act, and then watch and learn, as with any experiment*

– Healey (2018, p. 191)

**Simultaneous Cyber Operations (SCOs)** occur when multiple actors conduct cyber operations against each other continuously in cyberspace, and affecting, often degrading, the normal operation of the cyber system or enabled physical systems of each other. In this case we replace the attack rate $aI(t)$ by a constant attack rate (e.g. simply $a$).

The simplest, and most common, case is a dyadic interaction, with two adversaries, which we denote by the X-side and the Y-side, with persistent attacks at constant rates. It is important to note that unlike symbols representing mathematical variables, which are in italics, I use normal font for the two adversaries.

Instead of a single network as in the PAM (Figure 2.1), we now have two interacting networks (Figure 3.1). I will call this the **Fundamental Model of Simultaneous Cyber Operations (FMSCO)**. In Chapter 6, we will explore a variety of extensions of the FMSCO, so I hope you will be patient if your favorite modification is not included here.

## X-side assets

## Y-side assets

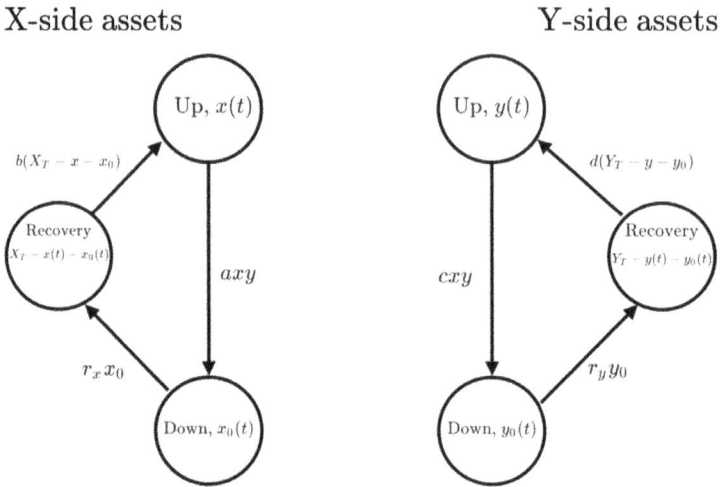

**Fig. 3.1.** In the Fundamental Model of Simultaneous Cyber Operations (FMSCO) two adversaries, the X-side and the Y-side, conduct persistent persistent attacks at constant rates. Instead of a single network as in the Pulse Attack Model (PAM), there are two interacting networks, which have similar dynamics for co-compromise and recovery as the PAM. We thus characterize each side by the number of uncompromised/up cyber assets $x(t), y(t)$, compromised/down cyber assets $x_0(t), y_0(t)$, and $x_r(t), y_r(t)$ recovering/resetting cyber assets at time $t$. When the total numbers of cyber assets $X_T, Y_T$ are constant, $x_r(t) = X_T - x(t) - x_0(t), y_r(t) = Y_T - y(t) - y_0(t))$. Note that unlike in the PAM, there is no co-compromise in the FMSCO (at least for now).

Before looking at the next section, try to formulate the dynamical equations for X-side and Y-side assets by yourself or in collaboration with a colleague just from the picture of the networks.

## 3.1.  Dynamics of the Cyber Assets

We denote by $X_T, x(t)$ and $x_0(t)$ the total number of X-side cyber assets, the number of uncompromised X-side cyber assets, and the number of compromised X-side cyber assets with similar symbology for the Y-side cyber assets. We also assume that cyber assets are neither permanently removed or added, so that the number of X-side cyber assets resetting at time $t$ is $x_r(t) = X_T - x(t) - x_0(t)$, with a similar expression for $y_r(t)$. We will not make use of $x_r(t)$ and $y_r(t)$ explicitly until extensions of the FMSCO in Chapter 6.

When the dynamics follow mass action

- Uncompromised X-side cyber assets decline due compromise by the Y-side at a rate proportional to the current numbers of uncompromised cyber assets on both sides (recall that there is no co-compromise) and increase at a rate proportional to the number that restored to operational status

$$\frac{dx}{dt} = -axy + b(X_T - x - x_0) \tag{3.1}$$

- Compromised X-side cyber assets increase due to compromise by the Y-side and decline as they are moved to resetting

$$\frac{dx_0}{dt} = axy - r_x x_0 \tag{3.2}$$

There is an analogous system of equations for the Y-side cyber assets:

$$\frac{dy}{dt} = -cxy + d(Y_T - y - y_0) \tag{3.3}$$

$$\frac{dy_0}{dt} = cxy - r_y y_0 \tag{3.4}$$

The rates $r_x$, $b$, $r_y$ and $d$ are key because they determine the ability of the X-side and Y-side to sustain cyber operations and determine performance of the cyber system or enabled physical system (e.g. Stanton and Tilton 2020).

The full system of four equations is rapidly solved numerically using deSolve, but easily visualizing the solution and understanding the dynamics of this four-dimensional system are not so simple and will occupy much of this chapter. In Table 3.1, I show the parameters for the base case computations.

## 3.2. Performance of the Cyber System or the Enabled Physical System

We continue to use Eqn. (1.4) to characterize the performance of the X-side cyber system or enabled physical system, and adopt a similar form for the performance of the Y-side cyber system or enabled physical system, so that when $y$ of the total $Y_T$ of the Y-side cyber

**Table 3.1.**   Variables, Parameters, and their Interpretation and Values for the FMSCO

| Symbol | Interpretation | Value in the base case |
|---|---|---|
| $X_T$ | Total number of X-side cyber assets | 40 |
| $Y_T$ | Total number of Y-side cyber assets | 120 |
| $a$ | Rate at which Y-side cyber assets compromise X-side cyber assets | 0.006 |
| $r_x$ | Rate at which compromised X-side cyber assets move to resetting | 0.1 |
| $b$ | Rate at which resetting X-side cyber assets return to operational status | 0.05 |
| $c$ | Rate at which X-side cyber assets compromise Y-side cyber assets | 0.0008 |
| $r_y$ | Rate at which compromised Y-side cyber assets move to resetting | 0.1 |
| $d$ | Rate at which compromised Y-side cyber assets return to operational status | 0.045 |

assets are uncompromised the performance functions are

$$\phi(x) = \left[ \frac{1}{1 + e^{\frac{x_{50} - x}{\sigma_x}}} \right], \quad \text{and}$$

$$\phi_y(y) = \left[ \frac{1}{1 + e^{\frac{y_{50} - y}{\sigma_y}}} \right] \tag{3.5}$$

Here the parameters $y_{50}$ and $\sigma_y$ have interpretation similar to those in the performance function for the X-side: $y_{50}$ is the number of uncompromised Y-side assets giving 50% performance and $\sigma_y$ determines how rapidly performance rises as the number of uncompromised Y-side cyber assets increases, as in Figure 1.6.

In Table 3.2, I show the parameters for the performance functions used in computations.

## 3.3.    Escalation to a Kinetic Attack or Cyber Attack on Critical Civilian Infrastructure

It is natural to envision cyber attacks in conjunction with kinetic attacks, and there are already examples of that. What about a cyber

**Table 3.2.** Functions and Parameters Characterizing Performance

| Symbol | Interpretation | Formula or Value |
|---|---|---|
| $\phi_x(x)$ | Performance of the X-side cyber system or the enabled physical system when $X(t) = x$ | Eqn. (1.4) |
| $\phi_y(y)$ | Performance of the Y-side cyber system or the enabled physical system when $Y(t) = y$ | Eqn. (3.5) |
| $x_{50}$ | Level of uncompromised X-side cyber assets giving 50% performance | 14 |
| $\sigma_x$ | Shape parameter for the X-side performance function | 4 |
| $y_{50}$ | Level of uncompromised Y-side cyber assets giving 50% performance | 65 |
| $\sigma_y$ | Shape parameter for the Y-side performance function | 15 |

attack leading to a kinetic response? When the actors are nations, the cyber security dilemma (Buchanan 2016, p. 20ff) has the sub-dilemmas of interpretation and response. With the former, a nation must determine the intentions of its adversary – how deeply will the persistent cyber operations go? This dilemma occurs in situations in which information is limited; Buchanan (2016, p. 96, 188) argues that nations will generally assume the worst about a cyber intrusion, so that escalation may occur not only during a crisis but in anticipation of one.

Clearly, the responses to an adversary's cyber operations vary on a spectrum of intensity but there is a major transition in response if a nation chooses to **escalate** to a kinetic attack or cyber attack on critical civilian infrastructure. Using cyber assets to cripple or destroy the cyber assets and/or the enabled physical system of an adversary has the potential to lead to either a physical (kinetic) attack or to an attack on critical civilian infrastructure such as hospitals, dams, and power systems. It is now generally agreed that international law applies in cyberspace and that nations should not destroy critical civilian infrastructure of other nations in peacetime (Buchanan 2016, pp. 134–135).

Reasons why escalation may not occur, in terms of both cyber and kinetic attacks, include (i) attribution is difficult (the attack may be done by rogue entities or third parties), increasing the possibility of mistakes in the response; (ii) if the victim has not prepared a list of potential targets of the adversary, responding may require too much time; and (iii) the cyberattack may be accidental or inadvertent—or its effects greatly exceeded the attacker's intentions (Libicki 2016, p. 328).

Even so, there is the real possibility of escalation from cyber to kinetic attacks, and issues include ethical considerations (Dipert 2010), counter-actions to reduce escalation (Kostyuk *et al.* 2018), and that the thresholds for such escalation are generally unknown to cyber adversaries (Geers 2010, Singer and Friedman 2014, Farrell and Glaser 2018, Rovner 2020). Arguments can be made that escalation to a kinetic attack is unlikely (Borghard and Lonergan 2019). On the other hand, the limits of coercion by cyber methods may increase the likelihood of attack on critical civilian infrastructure (Borghard and Lonergan 2017).

In cyber operations to date, adversaries behaved in ways to cause strategic effects but avoid escalation to armed conflict (Singer and Friedman 2014 Valeriano and Maness 2015, Warner 2020, Nakasone 2020). But this is not guaranteed for the future. Indeed, Singer and Friedman (2014, p. 136) quote an unnamed US military official saying "If you shut down our power grid, maybe we will put a missile down one of your smokestacks." On the other hand, Valeriano and Maness (2015, p. 54, 61ff) argue that the possibility of escalation to a kinetic attack or a cyber attack on critical infrastructure is one route to the emergence of restraint in cyber operations. Game extensions of the FMSCO (e.g Basar and Olsder 1982, Alpcan and Basar 2011, McNamara 2020) are beyond the scope of this book, but are likely a fruitful area of future research.

For these reasons, we will treat escalation as a probabilistic event, assuming that the goal of each nation during simultaneous cyber operations is to achieve the highest level of compromise of the adversary's assets without escalation by the adversary to a kinetic attack or a cyber attack on critical infrastructure (Smeets and Lin 2018).

For the Y-side attack on the X-side, we let $U_x(x, x_0)$ denote the probability that the X-side escalates when it has $x$ uncompromised cyber assets and $x_0$ compromised cyber assets. Understanding when an adversary will respond with a kinetic attack is involves the complicated matter of human behavior; we discuss this topic in Chapter 8. To treat escalation as a probabilistic event, we assume that the probability of escalation to a kinetic attack is determined by the absolute number of compromised assets $x_0$ and the relative performance $\frac{\phi_x(X_T)}{\phi_x(x)}$ of the X-side cyber system or enabled physical system when there are $x$ uncompromised cyber assets. We expect that the likelihood of escalation will increase as the number of compromised cyber assets increases and as performance decreases, which is why $\phi_x(x)$ is in the denominator in the previous expression. That is, because $\phi_x(X_T)$ is always greater than or equal to $\phi_x(x)$ the ratio $\phi_x(X_T)/\phi_x(x)$ increases as the number of uncompromised cyber assets decreases. To determine the probability of a kinetic response, we weight these by parameters the number of compromised cyber assets and relative performance by $\eta_{x_0}$ and $\eta_\phi$ respectively. We thus capture the X-side behavior as it depends upon the level of compromise and the reduction in performance of its cyber system or enabled physical system. For example, if $\eta_{x_0} = 0$, what matters to the X-side is the performance of its cyber system or enabled physical system, whereas if $\eta_\phi = 0$ what matters is the level of compromised assets, regardless of performance. In general, we expect that both of the weighting parameters will be non-zero.

For the mathematical form describing the probability of escalation, I use an exponential distribution (which we will discuss in more detail at the start of Chapter 4)

$$U_x(x, x_0) = 1 - e^{-\left[\eta_{x_0} x_0 + \eta_\phi\left(\frac{\phi_x(X_T)}{\phi_x(x)}\right)\right]} \tag{3.6}$$

> **Potential project**: Since $x_0 = 0$ when cyber operations commence, the first term in the exponent of Eqn. (3.6) is identically 0 and the ratio of the performance functions is 1. Thus, we require $\eta_\phi$ to be small for the probability of escalation to be small when SCOs start. In computations, I set $\eta_\phi = \eta_\phi = 0.02$. An alternative to Eqn. (3.6) allowing any choice of $\eta_\phi$ is $U_x(x, x_0) = 1 - e^{-[\eta_{x_0} x_0 + \eta_\phi(\frac{\phi_x(X_T)}{\phi_x(x)} - 1)]}$ and I encourage you to explore the differences between the two choices, both analytically and numerically.

We might ask: could the Y-side know the values of $x$ and $x_0$? For now, we will assume that the Y-side can monitor performance to infer a value for the number of uncompromised cyber assets and observed the X-side cyber system long enough to infer a value of $x_0$ and return to this question at the end of Chapter 5.

We define the value to the Y-side when attacking the X-side, $V_{yx}(x, x_0)$ as the reduction in the number of X-side cyber assets and performance of the X-side cyber system without escalation to a kinetic response by the X-side. When the X-side assets are $x$ and $x_0$, the probability of no kinetic response is $1 - U_x(x, x_0)$ and the reduction in performance is $\left(1 - \frac{\phi_x(x)}{\phi_x(X_T)}\right)$. The value to the Y-side in the attack is then

$$V_{yx}(x, x_0) = \left(1 - \frac{\phi_x(x)}{\phi_x(X_T)}\right)(1 - U_x(x, x_0)) \qquad (3.7)$$

There is an analogous set of equations for the value of the X-side attacking the Y-side (Table 3.3, which also includes parameters used for computations). As an exercise, I suggest that you write out a list of the parameters that determine $V_{yx}(x, x_0)$.

Intuition suggests that the X-side will be most likely to initiate a kinetic response or a cyber attack on critical civilian infrastructure when it has many compromised cyber assets and few uncompromised cyber assets, and this is the case (Figure 3.2, upper panel). I show the value for the Y-side when attacking the X-side in the lower panel of Figure 3.2. The parameters of the performance function for the X-side are such that the Y-side gains very little in terms of declining performance of the X-side until the number of uncompromised cyber assets is reduced below about 20. After that, there is a gradual increase in $V_{yx}(x, x_0)$, which rises to a peak and then declines. A figure such as Figure 3.2 allows decision-makers of the Y-side to select an attack rate with maximal effect, which we discuss in more detail in Chapter 6.

## 3.4.    Dynamics of the FMSCO: Numerical Results

We first consider the situation when simultaneous cyber operations initiate from no compromise on either side so that the initial conditions are $x(0) = X_T$, $x_0(0) = 0$, $y(0) = Y_T$, and $y_0(0) = 0$. We expect

**Table 3.3.** Functions and Parameters Characterizing Escalation

| Symbol | Interpretation | Formula or Value in the base case |
|---|---|---|
| $U_x(x, x_0)$ | Probability that the X-side escalates to a kinetic attack or an attack on critical infrastructure given $x$ uncompromised and $x_0$ compromised X-side cyber assets | Eqn. (3.6) |
| $U_y(y, y_0)$ | Probability that the Y-side escalates to a kinetic attack or an attack on critical infrastructure given $y$ uncompromised and $y_0$ compromised Y-side cyber assets | In analogy to Eqn. (3.6) |
| $\eta_x$ | Parameter characterizing the $x_0$ dependent increase in the probability that the X-side escalates to a kinetic attack or attack on critical civilian cyber infrastructure | 0.02 |
| $\eta_y$ | Parameter characterizing the $y_0$ dependent increase in the probability that the Y-side escalates to a kinetic attack or attack on critical civilian cyber infrastructure | 0.02 |
| $\psi_x$ | Parameter characterizing the performance dependent increase in the probability that the X-side escalates to a kinetic attack or attack on critical civilian cyber infrastructure | 0.02 |
| $\psi_y$ | Parameter characterizing the performance dependent increase in the probability that the Y-side escalates to a kinetic attack or attack on critical civilian cyber infrastructure | 0.02 |
| $V_{yx}(x, x_0)$ | The value of the Y-side attacking the X-side when there are to $x$ uncompromised and $x_0$ comprised X-side cyber assets | Eqn. (3.7) |
| $V_{xy}(y, y_0)$ | The value of the X-side attacking the Y-side when there are to $y$ uncompromised and $y_0$ comprised Y-side cyber assets | In analogy to Eqn. (3.7) |

that the number of uncompromised cyber assets will decline and the number of compromised cyber assets will increase. After exploring this case in detail, we consider initial conditions with different levels of compromised and compromised assets. As in Chapter 2, I used the RK4 option in the R package deSolve to solve Eqns. (3.1)–(3.4).

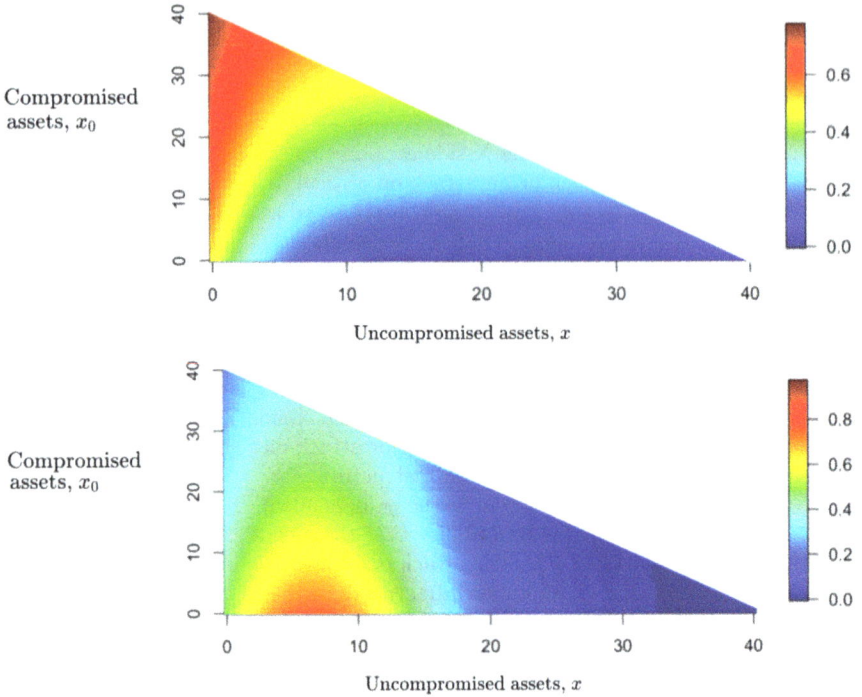

**Fig. 3.2.** Heat maps showing the probability that the X-side escalates to a kinetic attack or cyber attack on critical civilian infrastructure of the Y-side as a function of the number of uncompromised and compromised cyber assets from Eqn. (3.6) (upper panel) and the value to the Y-side of attack on the X-side from Eqn. (3.7) (lower panel).

### 3.4.1.  *Starting with only uncompromised cyber assets*

In Figures 3.3 and 3.4, I respectively show the dynamics of uncompromised and compromised cyber assets, the probabilities that either side escalates to a kinetic attack or a cyber attack on critical civilian infrastructure, and the value to each side of the attack on the other.

The numbers of uncompromised cyber assets of both sides decline monotonically in time towards steady state values (upper panel Figure 3.3) and the numbers of compromised cyber assets increase (lower panel, Figure 3.3), showing transient behavior before reaching the steady states. We denote the steady states by $\bar{x}$, $\bar{y}$, $\bar{x}_0$, and $\bar{y}_0$ and explore them in greater detail in Sections 3.5 and 3.6.

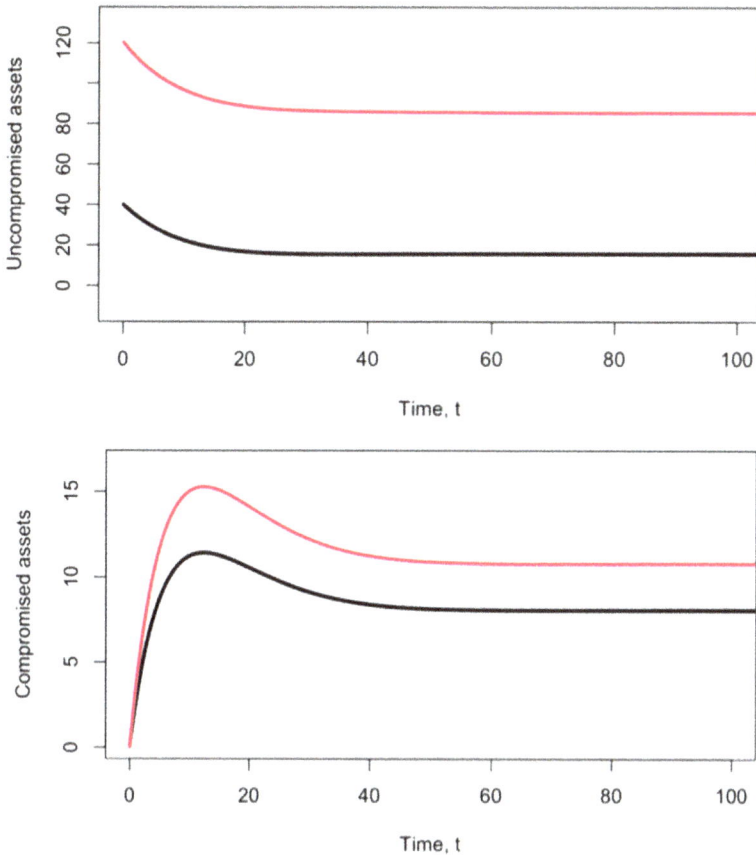

**Fig. 3.3.** The dynamics of cyber assets when simultaneous cyber operations are initiated with both sides having only uncompromised assets. Upper panel: The dynamics of X-side (black) and Y-side (red) uncompromised cyber assets. Lower panel: The dynamics of the corresponding compromised cyber assets. Note the different scales of the $y$-axes.

Because the probabilities of escalation depend upon both the numbers of compromised and uncompromised cyber assets, they too show a transient before reaching steady state values (Figure 3.4, upper panel). The transient behavior is less clear in the values of attack (Figure 3.4, lower panel).

As with the persistence of compromise in the PAM, the transient properties of compromise in the FMSCO are something to explain in the FMSCO.

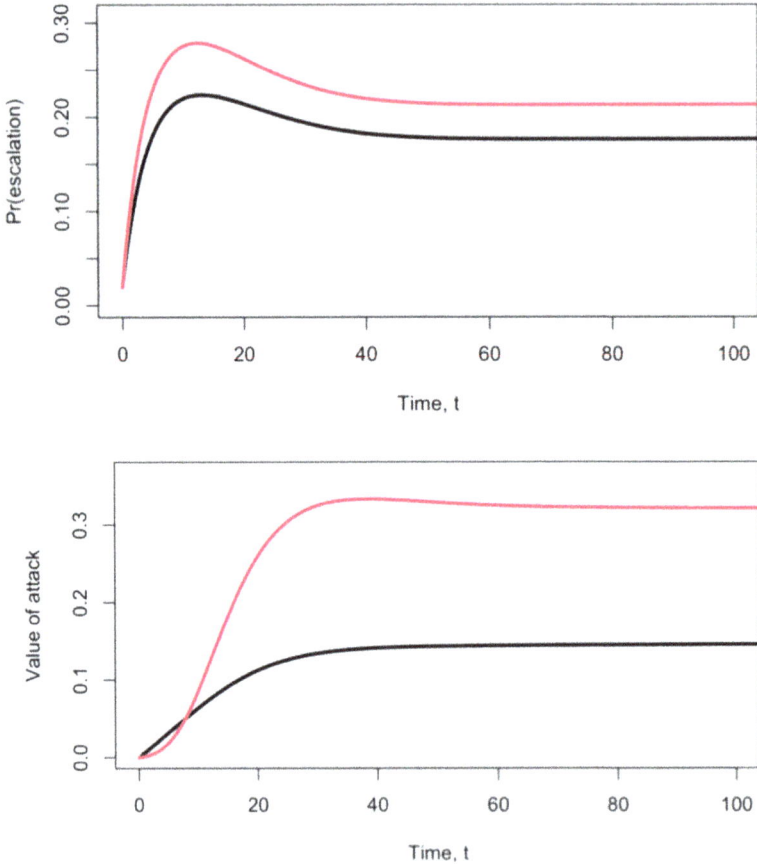

**Fig. 3.4.**   Upper panel: The probability that the X-side (black) or the Y-side (red) escalates to either a kinetic attack or a cyber attack on critical civilian infrastructure after simultaneous cyber operations are initiated. Lower panel: The values to the X-side of attacking the Y-side (black) and of the Y-side to attacking the X-side (red) after simultaneous cyber operations are initiated.

Furthermore:

- The steady state does not mean that some cyber assets are permanently compromised and others permanently uncompromised, but rather that cyber assets are constantly transitioning between compromised and uncompromised pools in a way that is balanced; see Mangel and Brown (2022) for additional examples and explanation.

- The mechanism leading to a positive steady state value of compromised assets in Figure 3.3 is very different than in Figure 2.2. In the lower panel of Figure 3.3 there is no co-compromise and a steady state of compromised cyber assets is reached because attack is continuous, while in Figure 2.2 the steady state of compromised cyber assets is reached even after offensive cyber operations have stopped because the co-compromise rate parameter $a_{co}$ exceeds the threshold value given in Eqn. (2.9).
- The steady state values in Figure 3.3 imply that one should expect that the performance of the cyber system or the enabled physical system permanently degrades. As discussed in the previous section, this may drive an increased risk of escalation to a kinetic attack or cyber attack on critical civilian infrastructure.
- The origin of the transient in the lower panel of Figure 3.3 will become clear in Section 3.6 but we can gain intuition with reference to Figure 3.1. When the X-side cyber assets are compromised, they go into the compromised pool from which there is waiting time, roughly $1/r_x$, before they move into the resetting/recovery pool, from which there is also a waiting time, roughly $1/b$, before the cyber assets are returned to uncompromised operational status. These waiting times lead to a pulse in compromised cyber assets, which ultimately settles down to a steady state as the flow of cyber assets into and out of the resetting pool match.

### 3.4.2. *Starting with a mixture of uncompromised, compromised, and resetting cyber assets*

The initial conditions could also be a mixture of uncompromised, compromised, and recovering cyber assets even if there is no previous cyber attack because natural variation can also lead to compromised assets (Mangel and Brown 2022). We can explore this idea by selecting a variety of values for $x(0)$ and $y(0)$ surrounding the steady state $(\overline{x}, \overline{y})$. These are the colored dots in Figure 3.5.

To do so, we specify either the initial number of compromised cyber assets or the initial number resetting cyber assets since choosing one of them and knowing the number of uncompromised cyber assets determines the other. Here, I specified the number of compromised cyber assets, so that the number of resetting cyber assets is then the difference between the total number of cyber assets

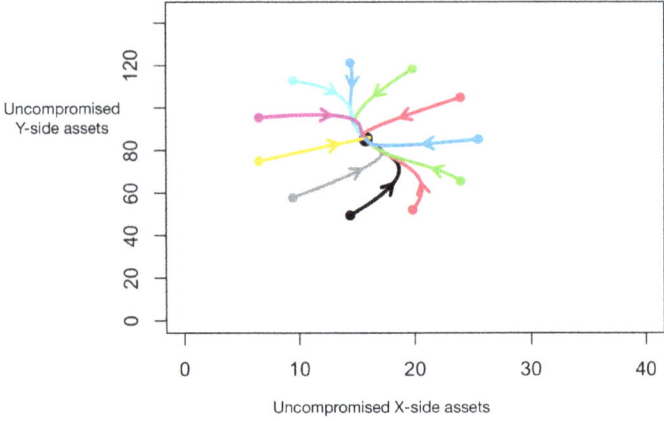

**Fig. 3.5.**    The solution of Eqns. (3.1)–(3.4) as trajectories in the phase plane of uncompromised X-side cyberassets/uncompromised Y-side cyber assets. The steady state of Eqns. (3.1)–(3.4) is the large black dot and the starting values for the number of uncompromised cyber assets $x(0)$ and $y(0)$ (the colored and small back dots) were chosen to surround the steady state. The starting values for compromised cyber assets were $x_0(0) = ax(0)y(0)/r_x$ and $y_0(0) = cx(0)y(0)/r_y$, as explained in the text. Each color corresponds to one of the trajectories in Figure 3.6. All trajectories flow towards the steady state.

and the sum of the compromised and uncompromised cyber assets. I chose $x_0(0)$ [and by analogy, $y_0(0)$] by noting that $dx_0/dt = 0$ in Eqn. (3.2) when the terms on the right side balance. Thus if we set $x_0(0) = ax(0)y(0)/r_x$ the initial conditions correspond to a momentary steady value for the number of compromised X-side assets. Similarly, I set $y_0(0) = cx(0)y(0)/r_y$. In choosing these initial conditions, we have to confirm that the choice does not exceed the total number of assets, i.e. that $x(0) + x_0(0) \leq X_T$ and $y(0) + y_0(0) \leq Y_T$.

In the next section, we will show that the steady state $(\overline{x}, \overline{y})$ is stable and unique, so that regardless of the starting point all trajectories go towards it (Figure 3.5). It appears that the trajectories cross, in contradiction of one of the fundamental properties of ordinary differential equations (Hartman 1973). This is perception caused by the projection of the four-dimensional solution $x/x_0/y/y_0$ into the $x/y$ plane. Were we to add the additional two variables ($x_0$ and $y_0$) and visualize in a four-dimensional space, the trajectories would not cross. Of course, in a three-dimensional world that is not possible, and we will subsequently explore ways to visualize the dynamics.

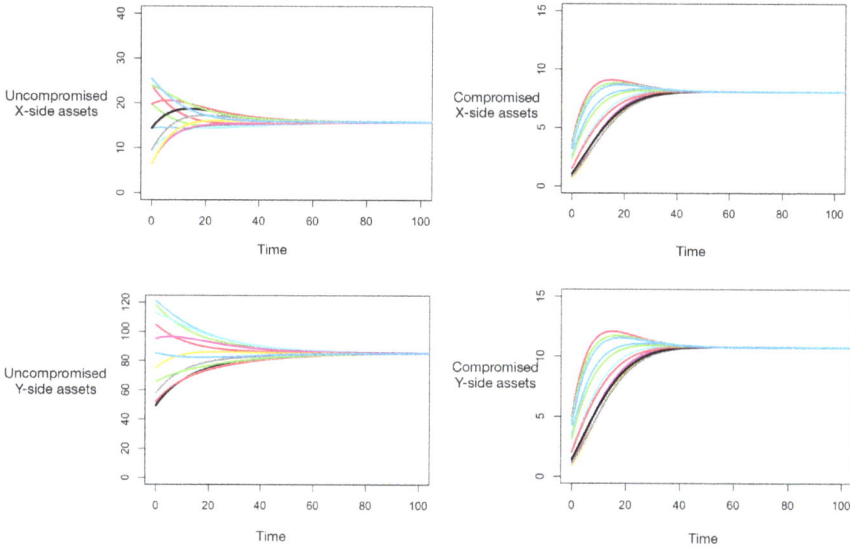

**Fig. 3.6.** The dynamics for X-side cyber assets (upper panels) and Y-side cyber assets (lower panels), determined from the solution of Eqns. (3.1)–(3.4). Each color represents the trajectory starting from one of the similarly colored dots in Figure 3.5. Regardless of the starting point, trajectories lead to the same steady state values.

We can also visualize the approach to the steady state as temporal dynamics, rather than the phase plane. Here (Figure 3.6) we see a smooth rise or decline from the initial value to the steady state of uncompromised assets and also a lack of transients as in the upper panel of Figure 3.3. This is due in part to the choice of initial conditions for the number of compromised assets. For example, if we set $x_0(0) = 0.25ax(0)y(0)/r_x$ and $y_0(0) = 0.25cx(0)y(0)$ some starting points lead to a smooth rise to the steady state, while others to a transient overshoot. The analysis in Sections 3.5 and 3.6 will allow us to understand the origin of this behavior.

### 3.4.3. *Design considerations: The resilience-performance tradeoff*

As in Chapter 2, we recognize that $b$ (Eqn. (3.1)) and $r_x$ (Eqn. (3.2)) are rates that represent operational capabilities of moving compromised cyber assets from the compromised pool to the resetting pool ($r_x$, Figure 3.1) and restoring cyber assets from the resetting pool to the uncompromised pool ($b$, Figure 3.1). Thus, we can envision a tradeoff between $b$ and $r_x$, in the sense that their sum is constrained. The most general form of a linear tradeoff would be $c_b b + c_r r_x$ is a constant. As in Chapter 2, I will set the constants $c_b = c_r = 1$ so that we do not need additional parameters, and in light of Table 3.1, for computations assume $b + r_x = 0.15$.

We now sweep over values of $b$, determining the dynamics of the cyber system and performance of the cyber system or the enabled physical system for a range of values of $b$, with $r_x$ determined by the constraint that $r_x = 0.15 - b$. Once the system reaches the steady state we compute steady state performance of the X-side assets from Eqn. (3.5). For computations, I let $b$ range from 0.015 to 0.135, divided into 50 equal values.

Intuition developed in Chapter 2 suggests that there will be an intermediate value of $b$ that maximizes performance. The reasoning is similar to that from Chapter 2: When $b$ is very large, once compromised cyber assets reach the resetting pool they are quickly repaired and sent back to uncompromised status. However, a large value of $b$ means that $r_x$ must small, so there will be a queue of cyber assets in the compromised pool waiting to be moved to the resetting/restoring pool. When $b$ is very small so that $r_x$ is large compromised cyber assets are moved quickly from the compromised pool to the resetting pool, but then a queue builds in the resetting pool, while cyber assets are metaphorically waiting for restoration and movement to the uncompromised pool. Our intuition about the intermediate value of $b$ is indeed the case (Figure 3.7, upper panel). The symmetric nature of steady state performance can be understood once we have a better grasp of the nature of the steady states (Section 3.5).

To compute resilience, imagine that once the steady state is reached, attack stops suddenly so that $a = 0$ in Eqns. (3.1) and (3.2). Because there is no co-compromise, the cyber system will fully recover, and we characterize resilience of the system by the time at which performance reaches a high fraction of the maximum steady state performance.

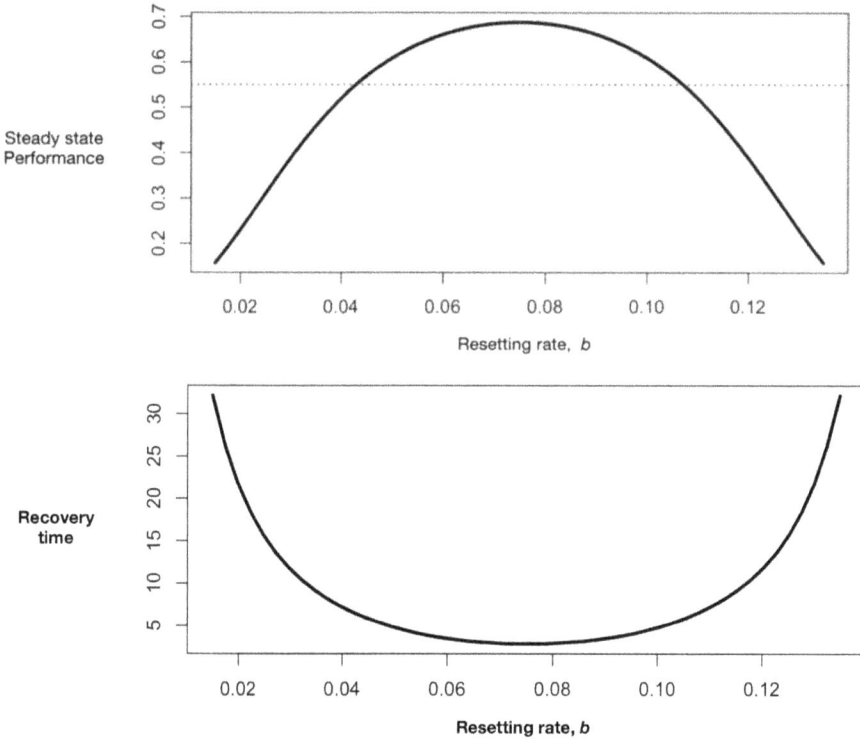

**Fig. 3.7.** An example of design considerations. Upper panel: Steady state performance of the X-side cyber system or the enabled physical system as the resetting rate $b$ varies when it and the rate $r_x$ at which compromised cyber assets are moved to resetting are constrained so that $b + r_x = 0.15$, which is the sum of their values in Table 3.1. The dotted line corresponds to steady state performance that is 80% of the maximum steady state performance. Lower panel: The recovery of performance after attack ends, measured by the time it takes performance to reach 80% of the maximum steady state performance.

Equations (3.1) and (3.2) become

$$\frac{dx}{dt} = b(X_T - x - x_0) \tag{3.8}$$

$$\frac{dx_0}{dt} = -r_x x_0 \tag{3.9}$$

so that the number of compromised assets $x_0(t)$ declines exponentially at rate $r_x$ and the number of uncompromised assets rises constantly towards $x(t) = X_T$. Performance will also increase in time, towards its maximum value determined by Eqn. (3.5). Suppose we define the resilience of the cyber system as the time it takes for performance to reach 80% of its maximum value). For the parameters in Tables 3.1 and 3.2, intermediate values of $b$ that give the fastest recovery in performance (Figure 3.7, lower panel). We conclude that for these parameters, there is minimal tradeoff between performance and resilience.

> **Potential project**: We have not yet addressed the possible tradeoff between cyber offense and defense (Slater 2017). That is, suppose that the attack rate parameter of the X-side, $c$, is also viewed as a resource and is incorporated into the resource constraint, as in $c_b b + c_r r_x + c_c c$, where now $c_b, c_r$ and $c_c$ are the unit costs of the resources $b, r_x$ and $c$. Explore this with your version of the FMSCO.

### 3.4.4.   *Analytical characterization of resilience\**

Equations (3.8) and (3.9) are linear first order differential equations, so we can obtain their explicit solutions. The solution of Eqn. (3.9) is $x_0(t) = \bar{x}_0 e^{-r_x t}$. We use this in Eqn. (3.8) and simultaneously move $bx$ to the left side so that Eqn. (3.8) becomes

$$\frac{dx}{dt} + bx = bX_T - b\bar{x}_0 e^{-r_x t} \tag{3.10}$$

We next note that

$$\frac{dx}{dt} + bx = e^{-bt}\frac{d}{dt}[xe^{bt}]$$

so that we can rewrite Eqn. (3.10) as

$$e^{-bt}\frac{d}{dt}[xe^{bt}] = bX_T - b\bar{x}_0 e^{-r_x t} \tag{3.11}$$

so that

$$\frac{d}{dt}[xe^{bt}] = bX_T e^{bt} - b\bar{x}_0 e^{-r_x t} e^{bt} = bX_T e^{bt} - b\bar{x}_0 e^{(b-r_x)t} \quad (3.12)$$

We integrate this equation from $t = 0$ when $x = \bar{x}$, to $t$ when the number of uncompromised cyber assets is $x(t)$. Doing so gives

$$x(t)e^{bt} - \bar{x} = X_T(e^{bt} - 1) - \frac{b\bar{x}_0}{b - r_x}(e^{(b-r_x)t} - 1) \quad (3.13)$$

Moving $\bar{x}$ to the right side and multiplying by $e^{-bt}$ gives

$$x(t) = e^{-bt}\bar{x} + X_T(1 - e^{-bt}) - \frac{b\bar{x}_0}{b - r_x}(e^{-r_x t} - e^{-bt}) \quad (3.14)$$

which we can rearrange as

$$x(t) = X_T + (\bar{x} - X_T)e^{-bt} - b\bar{x}_0\frac{e^{-r_x t} - e^{-bt}}{b - r_x} \quad (3.15)$$

Let us pause to interpret the three terms on the right side of Eqn. (3.15). When $t = 0$, the exponentials are equal to 1, so that the third term on the right side vanishes and $X_T$ on the first and second terms of the right side cancel, so that we have $x(0) = \bar{x}$, which is the assumption about the cyber assets of the X-side being in the steady state when the attack stops. Since always $\bar{x} - X_T < 0$, the second term on the right side is negative and declines towards 0 as time increases. What about the third term, particularly the fraction $\frac{e^{-r_x t} - e^{-bt}}{b - r_x}$, since we know that $b\bar{x}_0 > 0$? If $b > r_x$, the denominator of this fraction is positive, and the numerator is as well since $e^{-bt} < e^{-r_x t}$ and the fraction is positive. Similarly, if $b < r_x$ both the denominator and numerator are negative so that the fraction is positive. We thus conclude that because of the negative sign in front of the fraction, the third term on the right side is always a reduction in the number of uncompromised cyber assets.

We can then determine the resilience of the cyber system by setting $x(t)$ in Eqn. (3.15) to a large fraction of the total number of cyber assets, and finding the time at which this occurs, either graphically or by a numerical root solving process.

> **Potential project**: Our analysis will run into problems
> if $b = r_x$ since then the third term on the right side of
> Eqn. (3.15) is undefined. Although it is very unlikely that
> these two rates would be precisely equal, it is worth think-
> ing about them. When that $b = r_x$, Eqn. (3.12) becomes
>
> $$\frac{d}{dt}[xe^{bt}] = bX_T e^{bt} - b\overline{x}_0$$
>
> Solve this equation and plot its solution. Then let $r_x$ in
> Eqn. (3.15) be different from $b$ but close to it (e.g. 10% on
> either side), plot the solutions of Eqn. (3.15), and interpret
> the results. Finally, explore the properties of the fraction
> in Eqn. (3.15) either by Taylor expansion of the expo-
> nentials or application of L'Hoptial's rule of introductory
> calculus.

## 3.5.   Analysis of the Steady State

The nonlinearity in Eqns. (3.1)–(3.4) is very simple.[1] However, knowing that more complicated equations are ahead of us in subsequent chapters, we will not look for mathematically special solutions.
Instead, we will apply more general methods that have been used in
the study of the dynamics of populations for more than a century
(e.g. Bazykin 1998). The program that we follow is:

- Phase plane characterization of the steady states (Section 3.5.1).
- Perturbation analysis to characterize the nature of stability of the
  steady states (Section 3.5.2). This section has a $*$ next to it.
- Visualizing slices of the four-dimensional phase space (Section
  3.5.3).

These steps can be followed for any of the extensions in subsequent
chapters, so we work them out carefully in detail here. The future is

---

[1] That is, this nonlinearity is very simple because it involves only quadratic terms
that appear symmetrically in each of the four equations and one might be tempted
to look for special ways of solving these equations (Davis 1962) as we did with
the pulse attack model after the attack ended.

up to you. In accomplishing these steps, especially the phase plane and perturbation analysis, I will call on more advanced mathematical methods than we have used this far. A brief introduction to those methods can be found in Mangel (2006, Chapter 2), a more advanced treatment in Epstein (1997), and a thorough treatment in Edelstein-Keshet (1988), Murray (2002), or any junior-senior college level book on differential equations.

### 3.5.1. *Phase plane characterization of the steady states*

The steady states $\bar{x}, \bar{x}_0, \bar{y}$ and $\bar{y}_0$ obtained by setting the left sides of Eqns. (3.1)–(3.4) equal to 0 satisfy

$$a\bar{x} \cdot \bar{y} = b(X_T - \bar{x} - \bar{x}_0) \qquad (3.16)$$

$$a\bar{x} \cdot \bar{y} = r_x \bar{x}_0 \qquad (3.17)$$

$$c\bar{x} \cdot \bar{y} = d(Y_T - \bar{y} - \bar{y}_0) \qquad (3.18)$$

$$c\bar{x} \cdot \bar{y} = r_y \bar{y}_0 \qquad (3.19)$$

Equation (3.17) implies that $\bar{x}_0 = a\bar{x} \cdot \bar{y}/r_x$ so that Eqn. (3.16) becomes

$$a\bar{x} \cdot \bar{y} = b(X_T - \bar{x} - a\bar{x} \cdot \bar{y}/r_x) \qquad (3.20)$$

We solve this equation for $\bar{y}$ as a function of $\bar{x}$ by first rewriting it as

$$a\bar{x} \cdot \bar{y}[1 + b/r_x] = b(X_T - \bar{x}) \qquad (3.21)$$

Solving for $\bar{y}$ gives

$$\bar{y} = \frac{b(X_T - \bar{x})}{a\bar{x}(1 + b/r_x)} = \gamma\frac{X_T - \bar{x}}{\bar{x}} \qquad (3.22)$$

where $\gamma$ is the combination of parameters

$$\gamma = \frac{b}{a(1 + b/r_x)} \qquad (3.23)$$

We treat Eqns. (3.18) and (3.19) similarly and write

$$c\bar{x} \cdot \bar{y} = d(Y_T - \bar{y} - c\bar{x} \cdot \bar{y}/r_y) \qquad (3.24)$$

so that

$$c\bar{x} \cdot \bar{y}(1 + d/r_y) + d\bar{y} = dY_T \tag{3.25}$$

If we define $\delta = c(1 + d/r_y)$ then Eqn. (3.25) becomes

$$\bar{y} = \frac{dY_T}{\delta\bar{x} + d} \tag{3.26}$$

We can find $\bar{x}$ by setting the right sides of Eqns. (3.22) and (3.26) equal to each other, then $\bar{y}$ from either one of them, and then $\bar{x}_0$ and $\bar{y}_0$ follow directly from Eqns. (3.17) and (3.19); see Figure 3.8 for a graphical illustration.

Setting the right sides of Eqns. (3.22) and (3.26) equal to each other

$$\frac{dY_T}{\delta\bar{x} + d} = \gamma\frac{X_T - \bar{x}}{\bar{x}} \tag{3.27}$$

and simplifying gives the quadratic equation

$$\delta\bar{x}^2 + \bar{x}\left[\frac{dY_T}{\gamma} + d - \delta X_T\right] - dX_T \tag{3.28}$$

This quadratic equation has one positive and one negative solution (demonstrate that for yourself), so we conclude that the steady state of the FMSCO is unique. We will next explore its stability, both via perturbation analysis and visually.

### 3.5.2.   *Perturbation analysis to characterize the nature of stability of the steady state*

We now know that the steady state $(\bar{x}, \bar{x}_0, \bar{y}, \bar{y}_0)$ is unique and appears to be stable, as determined by numerical computation. We can gain more understanding of the steady state by conducting a formal stability analysis. We set

$$x(t) = \bar{x} + x_\epsilon(t)$$
$$x_0(t) = \bar{x}_0 + x_{0\epsilon}(t)$$
$$y(t) = \bar{y} + y_\epsilon(t)$$
$$y_0(t) = \bar{y}_0 + y_{0\epsilon}(t) \tag{3.29}$$

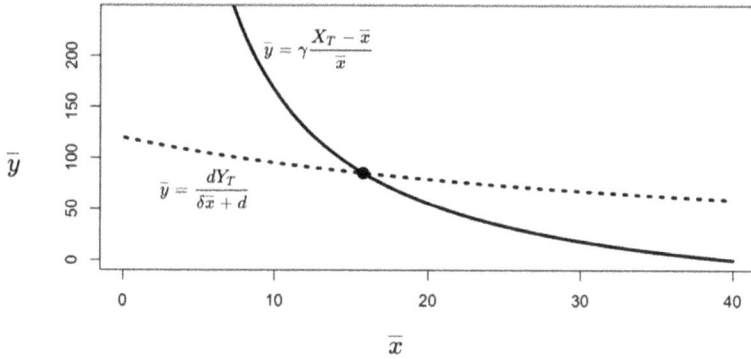

$$\bar{y} = \gamma \frac{X_T - \bar{x}}{\bar{x}}$$

$$\bar{y} = \frac{dY_T}{\delta \bar{x} + d}$$

**Fig. 3.8.** The steady state conditions in Eqns. (3.16) and (3.18) lead to two curves that relate $\bar{x}$ and $\bar{y}$ (Eqns. (3.22) and (3.26)). When these curves intersect, we determine $\bar{x}$ by setting the and $\bar{y}$, and then $\bar{x}_0$ and $\bar{y}_0$ follow directly from Eqns. (3.17) and (3.19). Thus, this is the steady state of the full four-dimensional system and visual inspection shows that they are the same values as those obtained by running the dynamics forward and shown in Figures 3.3–3.6.

where $\epsilon$ indicates a small number and the initial conditions for the second terms on the right-hand side of Eqn. (3.29) are understood to be proportional to $\epsilon$ so that we are starting the equations "not too far" from the steady state. Thus, for example, $x_\epsilon(0) = c_1\epsilon$, where $c_1$ is a constant that is of the order of 1, with the other initial conditions set similarly. We refer to any variable with a subscript $\epsilon$ as a perturbation term.

In the steady states Eqns. (3.1)–(3.4) become

$$\frac{dx_\epsilon}{dt} = 0 = -a(\bar{x} + x_\epsilon)(\bar{y} + y_\epsilon) + b(X_T - \bar{x} - x_\epsilon - \bar{x}_0 - x_{0\epsilon})$$

$$(3.30)$$

$$\frac{dx_{0\epsilon}}{dt} = 0 = a(\bar{x} + x_\epsilon)(\bar{y} + y_\epsilon) - r_x x_{0\epsilon} \tag{3.31}$$

$$\frac{dy_\epsilon}{dt} = 0 = -c(\bar{x} + x_\epsilon)(\bar{y} + y_\epsilon) + d(Y_T - \bar{y} - y_\epsilon - \bar{y}_0 - y_{0\epsilon}) \tag{3.32}$$

$$\frac{dy_{0\epsilon}}{dt} = 0 = c(\bar{x} + x_\epsilon)(\bar{y} + y_\epsilon) - r_y y_{0\epsilon} \tag{3.33}$$

We simplify Eqns. (3.30)–(3.33) by multiplying the right sides out, ignoring terms that are products of $x_\epsilon$, $y_\epsilon$, $x_{0\epsilon}$ and $y_{0\epsilon}$, because they are proportional to $\epsilon^2$, and apply the steady state conditions.

Equations (3.30)–(3.33) lead to a set of four linear equations for the dynamics of the perturbation terms

$$\frac{dx_\epsilon}{dt} = -(a\bar{y} + b)x_\epsilon - a\bar{x}y_\epsilon - bx_{0\epsilon} \tag{3.34}$$

$$\frac{dy_\epsilon}{dt} = -(c\bar{x} + d)y_\epsilon - c\bar{y}x_\epsilon - dy_{0\epsilon} \tag{3.35}$$

$$\frac{dx_{0\epsilon}}{dt} = a(\bar{x}y_\epsilon + \bar{y}x_\epsilon) - r_x x_{0\epsilon} \tag{3.36}$$

$$\frac{dy_{0\epsilon}}{dt} = c(\bar{y}x_\epsilon + \bar{x}y_\epsilon) - r_y y_{0\epsilon} \tag{3.37}$$

Using vector and matrix notation, we write Eqns. (3.34)–(3.37) as

$$\frac{d}{dt}\begin{pmatrix} x_\epsilon \\ y_\epsilon \\ x_{0\epsilon} \\ y_{0\epsilon} \end{pmatrix} = \begin{bmatrix} -(a\bar{y}+b) & -a\bar{x} & -b & 0 \\ -c\bar{y} & -(c\bar{x}+d) & 0 & -d \\ a\bar{y} & a\bar{x} & -r_x & 0 \\ c\bar{y} & c\bar{x} & 0 & -r_y \end{bmatrix} \times \begin{pmatrix} x_\epsilon \\ y_\epsilon \\ x_{0\epsilon} \\ y_{0\epsilon} \end{pmatrix} \tag{3.38}$$

The key to understanding the dynamics of the perturbation terms is to look at the eigenvalues and eigenvectors of the matrix, denoted by $\mathcal{M}$, on the right side of Eqn. (3.38). The eigenvalues and eigenvectors of $\mathcal{M}$ can be determined using the R command "ev ←eigen(M)" and then separately calling the eigenvalues or eigenvectors with the commands "(values ← ev$values)" and "(vectors ← ev$vectors)" respectively. Note that the matrix in Eqn. (3.38) is sufficiently simple that one could compute the eigenvectors by hand, but this will not be true in extensions of the FMSCO that we discuss in Chapter 6.

For the parameters in Table 3.1 the eigenvalues are

$$\lambda_1 = -0.100 \tag{3.39}$$

$$\lambda_2 = -0.0459 \tag{3.40}$$

$$\lambda_3 = -0.106 + 0.0556i \tag{3.41}$$

$$\lambda_4 = -0.106 - 0.0556i \tag{3.42}$$

Two of the eigenvalues are real and negative and the real part of the complex eigenvalues is also negative. This confirms that the numerical results that the steady state is locally stable. Since it is unique, it is also globally stable. The complex eigenvalues are conjugates of each other, so that there is a single period of oscillation associated with the decay to the steady state.

## 3.6. Looking at Slices of the Four-dimensional Phase Space

Equations (3.1)–(3.4) describe the dynamics of the two cyber systems in a four-dimensional phase space. This is hard to visualize! McPeek and colleagues (McPeek 2017, 2022, McPeek *et al.* 2022) suggested that we look at slices of the phase space to help visualize what is happening.

We can obtain a sense of how the cyber system moves from the initial state $x(0) = X_T, x_0(0) = 0, y(0) = Y_T, y_0(0) = 0$ towards the steady state by examining slices of the four-dimensional phase space in the four phase planes corresponding to (i) uncompromised X-side cyber assets/uncompromised Y-side cyber assets (upper left panel in Figure 3.9), (ii) uncompromised X-side cyber assets/compromised X-side cyber assets (upper right panel in Figure 3.9); (iii) uncompromised Y-side cyber assets/compromised Y-side cyber assets (lower left panel in Figure 3.9); and (iv) compromised X-side cyber assets/compromised Y-side cyber assets (lower right panel in Figure 3.9).

In the phase plane of uncompromised cyber assets (upper left panel Figure 3.9) by $t = 300$, uncompromised X-side cyber assets have dropped to less than half of their initial value, on the way to the steady state that is around 16; uncompromised Y-side cyber assets have dropped by 25% at the same time, on their way to the steady state that is around 85. Far from the steady state, the trajectory is almost linear, only showing nonlinear behavior near the steady state.

The trajectory in the the phase plane of X-side cyber assets (upper right panel in Figure 3.9) shows curvature from the outset, which accounts for the overshoot of compromised cyber assets shown in Figure 3.3. The trajectory in the phase plane of Y-side cyber assets (lower left panel in Figure 3.9) also shows curvature, similarly corresponding to the overshoot in the dynamics.

The trajectory in the phase plane for compromised cyber assets of both the X-side and Y-side (lower right panel in Figure 3.9) is

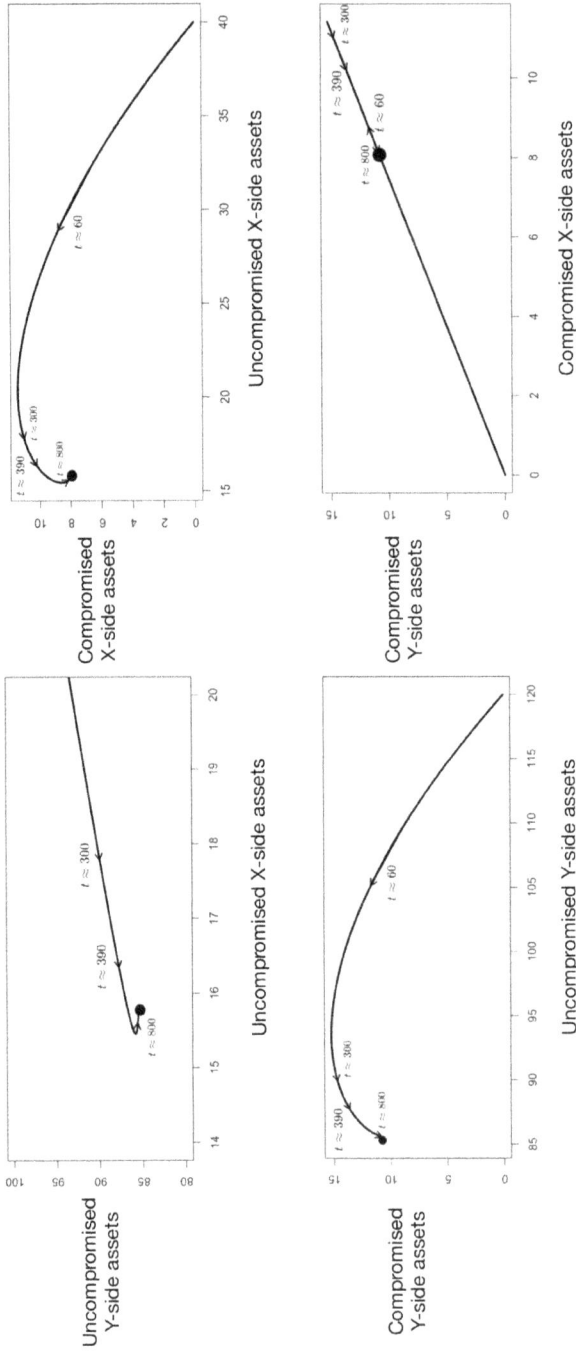

**Fig. 3.9.** Slices through the four-dimensional space when SCOs are initiated with only uncompromised cyber assets. In all panels, the arrows represent the direction in which cyber cyber assets move towards the steady state, shown by the black dot. Time is implicit in the trajectory, so I labeled various times along the trajectories. Upper left: The phase plane for uncompromised X-side and Y-side cyber assets. Since $x(0) = 40$ and $y(0) = 120$ we conclude that there is a relatively rapid decline in the number of uncompromised X-side cyber assets. Upper right: Phase plane of X-side cyber assets. Lower left: Phase plane of Y-side cyber assets. Lower right: Phase plane of X-side and Y-side compromised cyber assets. See text for details and interpretation.

perhaps the most interesting. Here, as in upper left panel, the trajectory in the phase plane is nearly linear, but the flow along that line is unlike the other three slices in the phase space because early on the flow is away from the steady state. Indeed, since the dynamics start at $x_0(0) = y_0(0) = 0$, it must be that in the full phase space the trajectory climbs above the $x_0/y_0$ plane and thus goes "over" the steady state $(\overline{x}_0, \overline{y}_0)$ while making a large excursion before "turning around" in the phase space and approaching the steady state.

We have learned a lot about the dynamics of the cyber system from these slices in the phase space. Sadly, I still do not know how to fully capture four-dimensional dynamics in a three-dimensional world.

### 3.6.1. *Summary of the analysis of the FMSCO*

Our combined analytical tools (computational and analytical) have shown that there is a unique, stable steady state for the dynamical equations characterizing the FMSCO. Approach to this steady state involves transient behavior of the numbers of uncompromised and compromised cyber assets, depending upon their values at the time that simultaneous cyber operations are initiated.

### 3.7. Extensions of the FMSCO

I hope that you have ideas about extending the FMSCO. We will do two kinds of extension in this book. The first extensions do not require new analysis, but rather applying what we have done thus far to modifications of the FMSCO. You can think of them as exercises or projects. The second extensions require additional new kinds of analyses; we will tackle many of them in Chapter 6.

For example, extensions that do not require new analysis, but rather applying what we have done thus far to are:

- *Including co-compromise in the FMSCO.* Perhaps the biggest difference between the PAM and the FMSCO is that the former includes co-compromise and the latter does not. In Chapter 6, we will modify Eqns. (3.1)–(3.4) to include co-compromise and discuss various aspects of these modified equations.

- *Using one's own cyber assets to hold the adversary's cyber assets in a compromised state.* When one side's cyber assets are used to hold the adversary's cyber assets in the compromised state (Long 2018) we need to consider an additional pool for the cyber assets. These are cyber assets that are uncompromised but committed to holding the adversary's cyber assets, which expands the number of differential equations in the FMSCO. We also do this in Chapter 6.

  Extensions that require new analysis are:

- *Stochastic versions of the PAM and FMSCO.* Ordinary differential equation models using mass action can be seen as a conditional mean of a stochastic population process (Gillespie and Mangel 1981, Mangel and Brown 2022). We will explain this in detail in Chapter 4. With stochastic models, the steady states of ordinary differential equation models are replaced by steady state distributions of outcomes. We will explore the connection between the ordinary differential equation and stochastic models, and gain new insights about cyber attack and defense.
- *Including a distribution of vulnerability to cyber attack.* In cyber systems with many kinds of cyber assets some cyber assets are likely more vulnerable to attack than others (Libicki 2018). One way to capture this is to modify the attack terms $axy$ and $cxy$ in Eqns. (3.1)–(3.4) so that the rates of attack have a temporal distribution that evolves as cyber attack occurs and more vulnerable assets are compromised. In Chapter 7, we will explore including a distribution of the rates of compromise in the PAM. I leave the further generalization of those ideas to the FMSCO to you.

## 3.8.    Summary of Major Insights

- During persistent cyber operations, both sides will experience permanent degradation in which the steady state of uncompromised cyber assets less than the initial number of uncompromised cyber assets is reached. This may correspond to a permanent degradation in performance of the cyber system or the enabled physical system, depending upon the parameters of the performance function and the number of uncompromised cyber assets in the steady state.

- The four-dimensional steady state of consisting of the uncompromised and compromised cyber assets of each side is unique and stable, meaning that regardless of the initial states of the cyber adversaries, the steady state will ultimately be approached. The approach to this steady state can include spiraling into it, rather than a monotonically approaching it.
- The value of the attack to the attacker can be determined from the probability of escalation by the adversary to a kinetic attack or a cyber attack on critical civilian infrastructure and the reduction in the performance of the adversary's cyber system or enabled physical system. These values also reach steady state values.
- Resilience, defined as the time to return to a near fully uncompromised cyber system if the cyber attack were to end, depends on both the rate at which compromised cyber assets are moved into the resetting pool and the rate at which they are moved from resetting to the uncompromised pool.
- It is a general property of simultaneous cyber operations that there are variables whose values shift from shrinking to growing or vice versa. An awareness of this property, as an anticipated and almost unavoidable phenomenon, is very important for decision-makers, who often pay great attention to trends but do not expect that they will reverse, or maybe even recognize that reversals are possible. Our analysis shows that a trend could shift direction without any outside influence.

## Chapter 4

# Beyond Determinism: Stochastic Versions of the PAM and FMSCO

*All the business of war, and indeed all the business of life, is to find out what you don't know by what you do; what I called 'guessing what was on the other side of the hill'*

– The Duke of Wellington

In titling this chapter, I riff off of the title of the book by Brian McCue *Beyond Lanchester* with subtitle *Stochastic Granular Attrition Combat Processes* (McCue 2020). There is a long history of using Ordinary Differential Equation (ODEs) models both in population biology (e.g. Gause 1934/2019) and military operations research (e.g. Lanchester 1917, Taylor 1983). At the same time, we recognize that fluctuations and stochasticity are common rather than rare in the natural world, so that there is a need to understand in what sense the deterministic ODEs are representative of the underlying stochastic processes that have more fidelity to the operational (or biological) situation.

Intuition suggests that the ODE models for populations are in some way representative of the mean of an underlying stochastic process and it is this connection that we want illuminate (e.g. Leslie and Gower 1958, Bartlett *et al.* 1960, Barnett 1962, Mangel and Ludwig 1977, Mangel 1994). McCue (2020) analyzes both Gause's classic experiments in population biology and classic military encounters such as the battle of Trafalgar, showing how distributions of outcomes arise. In **stochastic versions** of the PAM and the FMSCO, we introduce random components to the dynamics, so that instead of steady states and single numbers for quantities such as numbers of uncompromised cyber assets, performance, and recovery time, we obtain distributions.

To understand these extensions of the model in Chapters 2 and 3, you need to understand only the basic rules of probability and the **exponential distribution**, which we review in the context of search theory (also see Chapter 2 of Mangel 2006).

## 4.1.  Probability of Detection in Search

Washburn (1981/2014) conducted an experiment in which students (military officers who were doing advanced degrees at the Naval Postgraduate School) played games of search and detection against each other. This is clearly a situation in which intelligent agents were searching and evading, and yet when he analyzed the data, Washburn found that the empirical time to detection fit the exponential distribution (which underlies the stochastic versions of the ordinary differential equation models) very well (graph in the lower panel of Figure 4.1).

If the time to detection follows an exponential distribution with rate parameter $\lambda$

$$\Pr[\text{detection by time } t] = 1 - e^{-\lambda t} \tag{4.1}$$

so that the probability that the target remains undetected at time $t$ is $e^{-\lambda t}$. This leads to the very interesting "memoryless" property, understood in the following sense. Suppose that we are interested in $p(t, s) = \Pr[\text{no detection by time } t + s \text{ given no detection by time } t]$. We first recall the definition of conditional probability that $\Pr[\text{A given B}] = \Pr[\text{A and B}]/\Pr[\text{B}]$, where A and B are any two events with $\Pr[\text{B} > 0]$ (Feller 1968, Mangel 2006). In our case A is the event of no detection by time $t + s$ and B is the event of no detection by time $t$ so that event B is included in event A so that

$$p(t, s) = \frac{e^{-\lambda(t+s)}}{e^{-\lambda t}} = e^{-\lambda s} \tag{4.2}$$

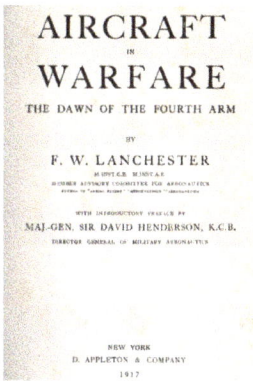

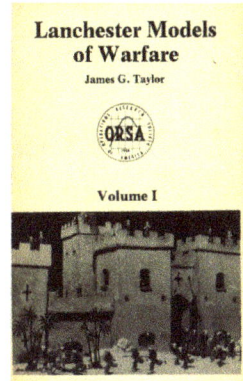

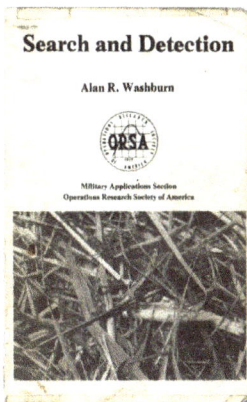

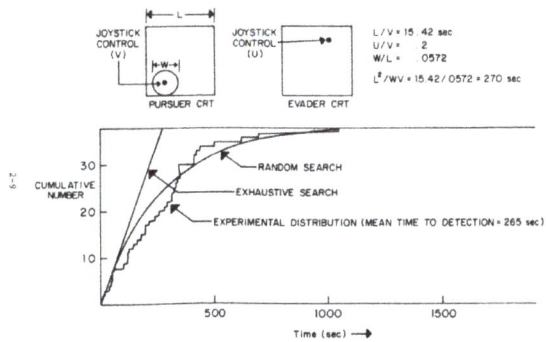

**Fig. 4.1.** Upper panel: the title page and frontispiece of Lanchester's 1917 book that pioneered the use of ordinary differential equations in the analysis of military operations and the cover of the first of the two volume classic by James G. Taylor (1983) in which those methods are extended. Lower panel: Cover of another classic, this one by Alan Washburn (1981/2014) on search theory, the set up for military officers playing games of search and detection, and the results in which the empirical data fit a distribution characterizing random search (see text for more details).

Thus, the failure of search up to time $t$ is "forgotten" as one goes forward in time; the search is called "random" (for further details, see Mangel 2006).

But surely, military officers playing this search and detection game were not forgetting what they had done! There are at least two explanations of what might be going on. The first (proposed by

Brian McCue) is that although each player remembers their actions, the other player – through movement of their own – eliminates the utility of that memory. The second is that there are other variables, which we may not know about, affecting utility of the memory.

There is something almost mystical in the way that simple mathematics can effectively capture complicated situations in nature. This was noted long ago by Eugene Wigner in a brilliant and accessible essay on mathematics and science (Wigner 1960).

Let us apply the concept of memoryless to the detection of cyber compromise by a defender. To do so, we assume that at time $t = 0$, the attacker has compromised a cyber asset and wishes to know the probability $P_U(t)$ that the compromise remains undetected at time $t$ in the future. We also assume that the probability that the defender detects compromise between time $t$ and $t + dt$, where $dt$ is a small increment of time, is $\lambda dt + o(dt)$ where $o(dt)$ represents terms that are higher powers of $dt$ such as $dt^2, dt^3$ and so forth. For compromise to remain undetected at time $t + dt$, it must be undetected at time $t$ and not detected in the interval between $t$ and $t + dt$. If we assume that detection of compromise is memoryless, these are independent events so that according to the rules of probability we multiply the probability of these events together giving

$$P_U(t + dt) = P_U(t)(1 - \lambda dt + o(dt)) \tag{4.3}$$

Simple algebra allows us to rearrange this equation to

$$\frac{P_U(t + dt) - P_U(t)}{dt} = -\lambda P_U(t) + \frac{o(dt)}{dt} \tag{4.4}$$

We now let $dt$ approach 0, in which case the left side of Eqn. (4.4) becomes the derivative $dP_U/dt$ and the higher-order terms $o(dt)/dt$ on the right side vanish, leading to

$$\frac{dP_U}{dt} = -\lambda P_U \tag{4.5}$$

and since compromise is undetected at $t = 0$, the solution is $P_U(t) = e^{-\lambda dt}$ so that corresponding probability of detection by time $t$ is $P_D(t) = 1 - e^{-\lambda t}$.

If this was your first experience of the Landau order notation $o(dt)$ and it is a bit mysterious, that is just fine because we will explore it further. And if the assumption that the rate of detection of compromise is constant troubles you, that is fine too because we will change that assumption in future chapters. We made that assumption here for its heuristic value and as a of warm up for the rest of this chapter.

## 4.2.    Stochastic Version of the Pulse Attack Model

To begin, we rewrite Eqns. (2.2) and (2.3) as difference equations. As above, we imagine a small interval of time $dt$ and let $dx = x(t + dt) - x(t)$, $dx_0 = x_0(t + dt) - x_0(t)$.

With this notation the ODEs for the PAM are equivalent to

$$dx = (-axI(t) - a_{co}xx_0 + b(X_T - x - x_0))dt + o(dt) \qquad (4.6)$$

$$dx_0 = (axI(t) + a_{co}xx_0 - r_xx_0)dt + o(dt) \qquad (4.7)$$

in the sense that if we divide both sides by $dt$ and then let $dt \to 0$ we obtain Eqns. (2.2) and (2.3).

Our goal is to go from these deterministic equations to a stochastic version that captures much of the same behavior in the mean but allows us to look at distributions, rather than point values, of states across time. Gillespie (1977, 2001, 2007) developed two kinds of methods for determining the probabilities of the possible changes in the states (here the numbers of uncompromised and compromised assets). The **Stochastic Simulation Algorithm** (SSA) fixes time steps and asks questions: (1) given the values of the state variables at time $t$, is there a change in them in the next $dt$ units of time and (2) if there is change, which state changes and by how much? The key here is to choose $dt$ sufficiently small that only one of the state variables changes, as is done in the derivation of the probability distribution for the Poisson process (Mangel 2006, pp. 95–100). The $\tau$-**Leaping Algorithm** also asks two questions (1) given the current values of the state variables, what is the time $\tau$ to the next change of any of the state variables and (2) what is the probability distribution of the change in the state variables? In this case there can be changes in more than one state between $t$ and $t + \tau$.

The SSA described is time-driven and the $\tau$-Leaping Algorithm is event driven. The latter is particularly handy when one is more interested in outcomes of interactions, such as the probability of success in a stochastic competition (Mangel and Ludwig 1977) or the outcome of stochastic versions of the Lanchester equations for battle (Mangel 1979, McCue 2020). I have used $\tau$-leaping in other work (Mangel and Bonsall 2008, Szekely *et al.* 2014); for our purposes the SSA is sufficiently speedy for computation (which is the great advantage of the $\tau$-Leaping Algorithm) and, I believe, allows clearer understanding of the relationship between the ordinary differential equations and the stochastic processes of interest.

### 4.2.1.   *The stochastic simulation algorithm*

In analogy to the PAM, we imagine a stochastic variables $X(t)$ [uncompromised cyber assets] and $X_0(t)$ [compromised cyber assets] and continue to assume that the relative rate of attack $I(t)$ is deterministic with the same structure as before. From the right-hand sides of Eqns. (4.6) and (4.7), we recognize three sub-processes affecting the dynamics of the states

- *Compromise* (*either by attack or co-compromise*) *of an uncompromised cyber asset.* This event decreases $X(t)$ by 1 (first two terms on the right-hand side of Eqn. (4.6)) and increases $X_0(t)$ by 1 (first two terms on the right-hand side of Eqn. (4.7)).
- *Return of a resetting cyber asset to the uncompromised state.* In this case $X(t)$ increases by 1 and $X_0(t)$ does not change (third term on the right-hand side of Eqn. (4.6)).
- *Removal of a compromised cyber asset to the resetting pool.* (third term on the right-hand side of Eqn. (4.7)) in which case $X(t)$ does not change and $X_0(t)$ decreases by 1.

We assume that the time interval $dt$ is so short that only one of these changes occurs.

We let $p_{\text{change}}(x, x_0, dt)$ denote the probability that the system changes in the next $dt$ units of time given $X(t) = x, X_0(t) = x_0$, where $x$ and $x_0$ are now generic values for the two random variables, not the solution of the ODEs. We assume that the probability of a change has an exponential distribution with rate parameter $\lambda(x, x_0)$ given by

$$\lambda(x, x_0) = axI(t) + a_{co}xx_0 + b(X_T - x - x_0) + r_x x_0 \qquad (4.8)$$

That is, in the SSA the rate of the transition process is determined by summing the right hand sides of the ODEs motivating the model, treating every term as positive. The probability of no change is then

$$p_{\text{nochange}}(x, x_0, dt) = e^{-\lambda(x,x_0)dt} \qquad (4.9)$$

If there is no change, $X(t + dt) = X(t)$ and $X_0(t + dt) = X_0(t)$. The probability of change in the next $dt$ units of time is

$$p_{\text{change}}(x, x_0, dt) = 1 - e^{-\lambda(x,x_0)dt} \qquad (4.10)$$

and there are three possible transitions. Letting $dX = X(t + dt) - X(t)$ and $dX_0 = X_0(t + dt) - X_0(t)$ we can denote these transitions by

$$\mathcal{T}_1 = \{dX = -1, dX_0 = 1\}$$
$$\mathcal{T}_2 = \{dX = 1, dX_0 = 0\}$$
$$\mathcal{T}_3 = \{dX = 0, dX_0 = -1\} \qquad (4.11)$$

Given that a change has occurred, we next assign probabilities $p_i$ to the transitions $\mathcal{T}_i$ for $i = 1, 2, 3$. With reference to the ODES in Eqns. (4.6) and (4.7), the rate of the transition $\mathcal{T}_1$ is $\mathcal{R}_1 = axI(t) + a_{co}xx_0$, of $\mathcal{T}_2$ is $\mathcal{R}_2 = b(X_T - x - x_0)$, and of $\mathcal{T}_3$ is $\mathcal{R}_3 = r_x x_0$. These rates sum to $\lambda(x, x_0)$, as they must because they capture all that could take place. We define the probability of the different transitions as the relative contributions of each of the transitions to $\lambda(x, x_0)$

$$p_1(x, x_0) = \frac{axI(t) + a_{co}xx_0}{axI(t) + a_{co}xx_0 + b(X_T - x - x_0) + r_x x_0}$$

$$p_2(x, x_0) = \frac{b(X_T - x - x_0)}{axI(t) + a_{co}xx_0 + b(X_T - x - x_0) + r_x x_0}$$

$$p_3(x, x_0) = \frac{r_x x_0}{axI(t) + a_{co}xx_0 + b(X_T - x - x_0) + r_x x_0} \qquad (4.12)$$

With these definitions, we ensure that the sum of the $p_i(x, x_0)$ is 1.

When a change occurs, we use Eqns. (4.10)–(4.12) to determine which transition occurred; when it is transition $i$ we have

$$\{X(t + dt), X_0(t + dt)\} = \{X(t), X_0(t)\} + \mathcal{T}_i \qquad (4.13)$$

where we add the components of these vectors individually.

To determine which transition occurs and evolution of the stochastic process $\{X(t), X_0(t)\}$, we use Monte Carlo simulation.

### 4.2.2.  *Why not stochastic differential equations?*[*]

More mathematically inclined readers may wonder why we are using a discrete time stochastic process instead of Stochastic Differential Equations (SDEs). It is a worthwhile question. That is, why not replace Eqns. (2.2) and (2.3) by

$$dX = [-axI(t) - a_{co}XX_0 + b(X_T - X - X_0)]dt + \sigma_1(X, X_0)dB_1$$

$$dX_0 = [aXI(t) + a_{co}X_0 - r_x X_0]dt + \sigma_2(X, X_0)dB_2$$

where $dB_1$ and $dB_2$ are increments in standard Brownian motion and $\sigma_1(X, X_0)$ and $\sigma_2(X, X_0)$ are the standard deviations attached to the Brownian increments.

If you are not one of the more mathematically inclined readers but would like to know about standard Brownian motion, introductions are given in Mangel (1985) and Mangel (2006); a more advanced but accessible treatment is in Thygessen (2023).

In brief, there are three reasons why SDEs are beyond the scope of this book:

- *Philosophical* Which stochastic calculus is applicable? SDEs are generally interpreted following Ito or Stratonovich, depending upon how one views the correlation between increments in the $dB_i$. But Krener (1979) showed that there are many more possible interpretations of the relevant stochastic calculus.
- *Mechanism* We can be pretty certain that the $\sigma_i$ are not constants, but depend upon the state variables, which is why I wrote them as $\sigma_1(X, X_0)$ and $\sigma_2(X, X_0)$. One way to determine them is to return to the underlying birth and death process and use it to model the choice of the $\sigma_i$, as in the papers of Leslie and Gower (1958), Bartlett *et al.* (1960), Barnett (1962), and Mangel and Ludwig (1977). Thus, we would have to wrestle with these processes in order to make sense of the parametrization of the SDEs.
- *Mathematical complexity* One advantage of the SDE approach is that it allows one to describe the probability density of uncompromised and compromised assets and recovery times in terms of

partial differential equations. For even the simplest problem that we study, these are at least three-dimensional ($t, x$, and $x_0$) partial differential equations that require either approximate solutions (e.g. Mangel and Ludwig 1977) or tricky numerical schemes. Doing so is beyond the level of this book.

If you are a lover of SDEs (as am I), these reasons are not meant to discourage you, but rather to encourage future work on stochastic version of the PAM and FMSCO, but we will stick with the discrete time stochastic process.

### 4.2.3. *Implementation of the stochastic simulation algorithm*

For numerical implementation, I use the same values for $a, a_{co}, b, r_x$ and $T$ as in Chapter 2.

The first decision is to choose the time interval $dt$ for use in Eqns. (4.10) and (4.13). That is, we divide the time interval of interest $[0,T]$ into $N$ pieces, so that $dt = T/N$. Recall that when we solved the ODES in Chapter 2, we used the fourth order Runge Kutta numerical scheme included in the package deSolve. That scheme is fourth order, which means that the error in iterating from one time step to the next is $o(dt^4)$ (Abramowitz and Stegun 1965, p. 896). For the results in Chapter 2, I used $dt = 0.05$, so that the error was $o(0.05^4) = 6.25 \ 10^{-6}$. However, Eqn. (4.13) – the iteration of the stochastic process – is like an Euler numerical scheme, so that it has error $o(dt^2)$. Clearly, the same choice of $dt = 0.05$ will not provide the accuracy we need for the stochastic simulation. But we need to remember that the smaller $dt$ is, the longer the simulation will take to run, so setting $dt = 0.05^3 = 0.000125$ may not be advisable, For the results in this section, I set $dt = 0.00625$ and we will see that it is a sufficiently small time interval, in the sense that the mean of the stochastic trajectories matches the deterministic one.

Here is pseudo-code to implement the SSA;

- **Step 1:** Import the parameters from the ODEs used in Chapter 2, setting $dt = 0.00625$. This determines the time-like dimension of the stochastic process by setting $N = T/dt$. Set the number of iterations of the simulation, $S$. For results shown in the next section, I set $S = 500$. To ensure that the pulse attack has the new dimension $N$, recompute it with the new value of $dt$ and let $I[n]$ denote the relative intensity of the pulse at time step $n$.

- **Step 2:** Dimension and initialize the stochastic process, using the subscript "sim" denote simulation variables. The simulated stochastic process has dimensions $X_{\text{sim}}[S, N]$ and $X_{0,\text{sim}}[S, N]$. That is, $X_{\text{sim}}[s, n]$ is the number of uncompromised assets in simulation $s$ at time step $n$. Assuming that the initial condition is a fully uncompromised cyber system (as is appropriate for the pulse attack), for every $s = 1, \ldots, S$ set $X_{\text{sim}}[s, 1] = X_T$ and $X_{0,\text{sim}}[s, 1] = 0$.

- **Step 3:** As the first check on the choice of $dt$, iterate forward assuming that the changes in the stochastic process always follow the mean changes of $X(t)$ and $X_0(t)$. Since this will be a deterministic computation, we need to do it only once and I drop the variable $s$ in this step. Thus we cycle from $n = 1$ to $n = N$, use the notation $E_{dX}$ and $E_{dX_0}$ to denote the mean changes in $X(t)$ and $X_0(t)$ and write

$$
\begin{aligned}
E_{dX} = {} & dt \cdot (-aX_{\text{sim}}[n] \cdot X_{0,\text{sim}}[n] - a_{co} \cdot X_{\text{sim}}[n] \cdot X_{0,\text{sim}}[n] \\
& + b \cdot (X_T - X_{\text{sim}}[n] - X_{0,\text{sim}}[n]) \\
E_{dX_0} = {} & dt \cdot (aX_{\text{sim}}[n] \cdot X_{0,\text{sim}}[n] + a_{co} \cdot X_{\text{sim}}[n] \cdot X_{0,\text{sim}}[n] \\
& - r_x X_{\text{sim}}[n]) \\
X_{\text{sim}}[n+1] = {} & X_{\text{sim}}[n] + E_{dX} \\
X_{0,\text{sim}}[n+1] = {} & X_{0,\text{sim}}[n] + E_{dX_0}
\end{aligned}
$$

(when coding this step we replace $\cdot$ by $*$, but these are equations, not computer code, and good writing practice calls on us to use the dot rather than the asterisk). If we have done a good job choosing $dt$, the trajectories generated in this step should be close to the solution of the ODE.

- **Step 4:** We are now ready to implement the full SSA. In a computer code we cannot generally use mathematical symbols, so I will use text for symbols in the equations that are above. For example, in both my code and pseudocode instead of using $\lambda$, I write *lambda*. We cycle over simulations, i.e. from $s = 1$ to $s = S$. For each simulation, we cycle over time, i.e. from $n = 1$ to $N - 1$. At each $n$ we first compute the transition rate (Eqn. (4.8))

$$lambda = a \cdot X_{\text{sim}}[s, n] \cdot I[n] + a_{co} \cdot X_{\text{sim}}[s, n] \cdot X_{0,\text{sim}}[s, n]$$
$$+ b(X_T - X_{\text{sim}}[s, n] - X_{0,\text{sim}}[s, n]) + r_x \cdot X_{0,\text{sim}}[s, n]$$

and from this determine whether the values of the states change or not. Letting $p_{nc}$ denote the probability of no change, from Eqn. (4.10) we have

$$p_{nc} = \exp(-lambda \cdot dt)$$

We now draw a random number uniformly distributed between 0 and 1; let's denote it by $Z$ (e.g. the command in $R$ is $Z = \text{runif}(1, 0, 1)$). If $Z < p_{nc}$ then no change occurs; go to Step 5. If $Z >= p_{nc}$ then a change occurs in the stochastic process; go to Step 6.

- **Step 5:** Since no change occurs

$$X_{\text{sim}}[s, n + 1] = X_{\text{sim}}[s, n]$$
$$X_{0,\text{sim}}[s, n + 1] = X_{0,\text{sim}}[s, n]$$

We then increment $n$ to $n + 1$. If $n + 1 < N$, we return to Step 4 in the same replicate of the simulation. If $n + 1 = N$ and $s < S$ we increment $s$ to $s + 1$ and return to Step 4 in the next replicate of the simulation. If neither of these situations holds, we are done with the simulation and ready to visualize results (go to Step 7).

- **Step 6:** We reach this step when a change occurs in the system, so need to determine which of the three transitions occurs using Eqn. (4.12) by setting

$$p_1 = \frac{a \cdot X_{\text{sim}}[s, n] \cdot I[n] + a_{co} \cdot X_{\text{sim}}[s, n] \cdot X_{0,\text{sim}}[s, n]}{lambda}$$

$$p_2 = \frac{b(X_T - X_{\text{sim}}[s, n] - X_{0,\text{sim}}[s, n])}{lambda}$$

$$p_3 = \frac{r_x X_{0,\text{sim}}[s, n]}{lambda}$$

Now draw another random number uniformly distributed between 0 and 1, continuing to use the symbol $Z$, and with it determine which of the transitions occur. If $Z < p_1$ then transition 1 occurs so that

$$X_{\text{sim}}[s, n+1] = X_{\text{sim}}[s, n] - 1$$
$$X_{0,\text{sim}}[s, n+1] = X_{0,\text{sim}}[s, n] + 1$$

If If $Z \geq p_1$ and $Z < p_1 + p_2$ then transition 2 occurs so that

$$X_{\text{sim}}[s, n+1] = X_{\text{sim}}[s, n] + 1$$
$$X_{0,\text{sim}}[s, n+1] = X_{0,\text{sim}}[s, n]$$

Finally, if $Z \geq p_1 + p_2$ then transition 3 occurs so that

$$X_{\text{sim}}[s, n+1] = X_{\text{sim}}[s, n]$$
$$X_{0,\text{sim}}[s, n+1] = X_{0,\text{sim}}[s, n] - 1$$

As in Step 5, we then increment $n$ to $n+1$. If $n+1 < N$, we return to Step 4 in the same replicate of the simulation. If $n+1 = N$ and $s < S$ we increment $s$ to $s+1$ and return to Step 4 in the next replicate of the simulation. If neither of these situations hold, we are done with the simulation and ready to visualize results (go to Step 7).

- **Step 7:** Our simulation is complete, and we are ready to visualize. I will not write pseudocode for that.

## 4.3.   Results for the Stochastic PAM

Since qualitative behavior of the differential equations depends upon whether the co-compromise rate parameter $a_{co}$ exceeds the threshold for persistence $r/X_T$ of compromise or not, we consider the results in two sub-sections. In each case we explore the dynamics of compromise and the distribution of the minimum value of the number of uncompromised cyber assets, and recovery times when $a_{co}$ is less than $r/X_T$ or the quasi-steady state distribution of the number of compromised assets when $a_{co}$ is greater than $r/X_T$. Before reading on, think a bit about what you expect might happen in the stochastic model.

### 4.3.1. $a_{co}$ is less than the threshold value for the persistence of compromise

The upper panel in Figure 4.2, reproduces the results from Chapter 2, using the ODE model, with the addition of the time at which the number of uncompromised assets return to 98% of $X_T$ (vertical dotted line). As before the blue dotted line is the pulse attack, the black line is the number of uncompromised cyber assets and the red line is the number of compromised cyber assets. The middle panel shows the trajectory of uncompromised cyber assets (black), identical to the black trajectory in the upper panel) and the trajectory of the conditional mean in the SSA (Step 3). In this panel, I have offset time by 1 unit because otherwise the curves sit on top of each other. The lower panel shows the first 10 trajectories of the SSA. All trajectories are identical before attack starts and long after attack ends because in this case compromise does not persist (i.e. a single red line) but during attack there is variation in the number of uncompromised cyber assets. It is this variation that interests us.

Two ways of capturing the variability in compromise and recovery are shown in Figure 4.3. In the upper panel, I show the distribution of the minimum number of uncompromised cyber assets, which varies from about 450 to about 550 with a peak around 500. That is, there is a 20% variation in the minimum number of uncompromised cyber assets simply due to natural fluctuations.

In the lower panel, there is similar variation in the time to return to $0.98X_T$ (i.e. a range from about 95 to 115 with a peak around 100). In this case, the distribution is much less symmetrical than that for the number of uncompromised cyber assets.

Remember that there is no difference in the underlying parameters, so that all of the variation in Figure 4.3 is caused by fluctuations in the underlying birth and death processes. From this observation, we learn that one needs to careful in asserting that the trajectory with a minimum of 450 uncompromised cyber assets is more poorly defended than one with a minimum of 550 uncompromised cyber assets, or that recovery that takes until $t = 115$ is more poorly done than recovery that takes until $t = 95$. We should not let stochasticity mislead us.

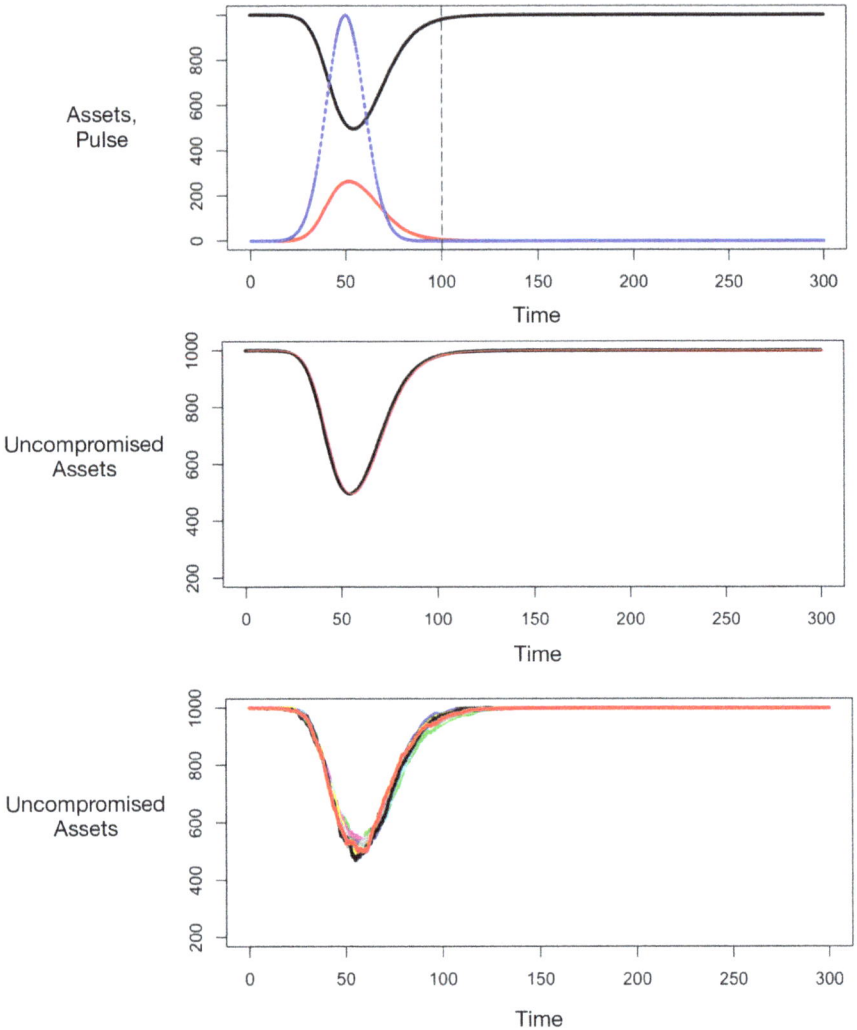

**Fig. 4.2.** Upper panel: The dynamics of the ODE model from Chapter 2 [the vertical dotted line corresponds to the time at which the number of uncompromised cyber assets reaches 0.98 $X_T$] when $a_{co}$ is less than the threshold value for the persistence of compromise. Middle panel: the number of uncompromised cyber assets from the ODE model (black) and the conditional mean trajectory computed in Step 3 of the SSA (red) for the PAM [the trajectory from the ODEs is plotted at time corresponding to $n + 1$ so that the curves do not sit on top of each other]. Lower panel: the first 10 trajectories from the SSA for the PAM when $a_{co}$ is less than the threshold value in the lower panel.

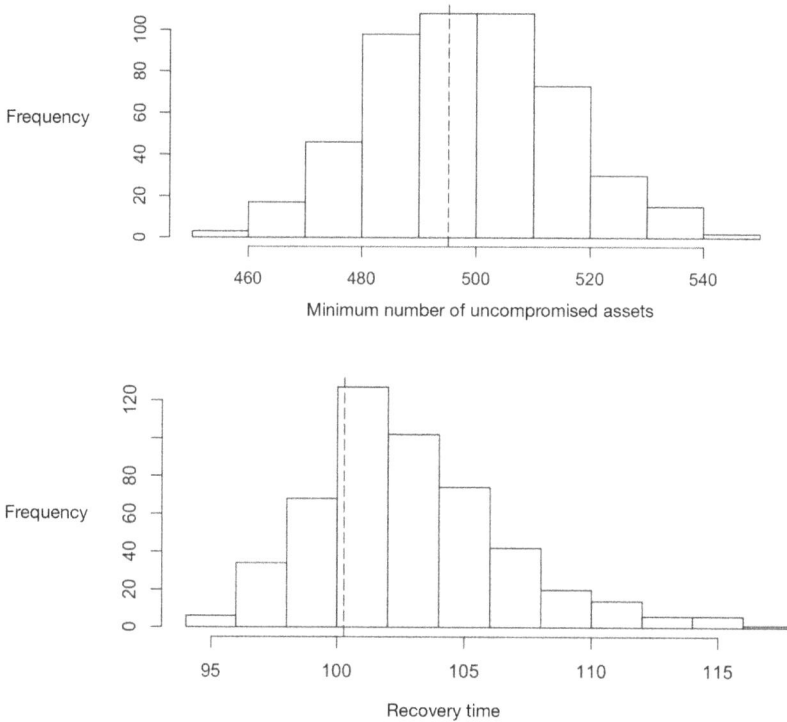

**Fig. 4.3.**  We compare the solution of the ODEs and the results of the SSA for the PAM when $a_{co}$ is less than the threshold value for the persistence of compromise by considering the distribution of the minimum value of the number of uncompromised cyber assets (upper panel) and the time to recover to $0.98X_T$ (lower panel). The vertical dotted lines show the values obtained from the ODES.

### 4.3.2.    $a_{co}$ exceeds the threshold value for the persistence of compromise

Figure 4.4 is the analogue of Figure 4.2 for the situation in which compromise persists. Here the dynamics are more interesting. For example, compare the middle panels of the two figures. In addition to the persistence of compromise the conditional mean trajectory computed in Step 3 of the SSA shows slow oscillations.

Now compare the lower panels in Figures 4.2 and 4.4. In Figure 4.2, a long enough time after the attack ends, compromise has disappeared and all simulated trajectories converge. On the other hand, in Figure 4.4, because compromise persists, we continue to see

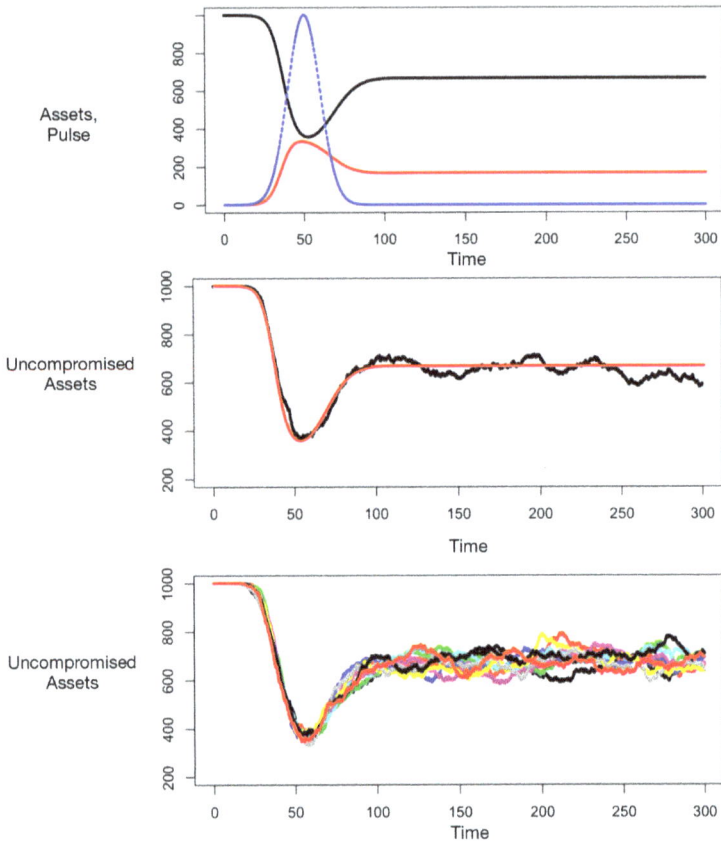

**Fig. 4.4.** The dynamics when $a_{co}$ exceeds the threshold value for the persistence of compromise in the PAM. Upper panel: ODE model from Chapter 2 (as before the pulse attack is shown in blue, uncompromised cyber assets in black, and compromised assets in red). Middle panel: the number of uncompromised cyber assets from the ODE model (black) and the conditional mean trajectory computed in Step 3 of the SSA (red) [in which the trajectory from the ODEs is plotted at time corresponding to $n + 1$ so that the curves do not sit on top of each other]. Lower panel: the first 10 trajectories from the remaining steps in the SSA for the PAM.

fluctuations around the deterministic level of long term compromise. That is, in this case, the cyber system will not reach a steady state but continue to have properties that are "random-walk" like. In a beautiful (but mathematically sophisticated) paper, Parsons (2018) shows that this kind of behavior is expected when the original ODE system has logistic-like properties, that is, beginning at small values, population numbers rise rapidly at first and then settle down to a steady state. If you are one of the more mathematically inclined readers, taking a look at Parsons's paper is well worth it.

Figure 4.5 is the analogue of Figure 4.3. The upper panel – concerning the minimum value of uncompromised cyber assets – tells the same story as the upper panel in Figure 4.3, but with more pronounced effect: now the range is from about 300 to 400 uncompromised cyber assets, with a peak around 350, so that the range is closer to 30%.

We can no longer discuss recovery time, as in the lower panel in Figure 4.3, but we can discuss the distribution of uncompromised cyber assets at the terminal time $T$, which we can interpret as the

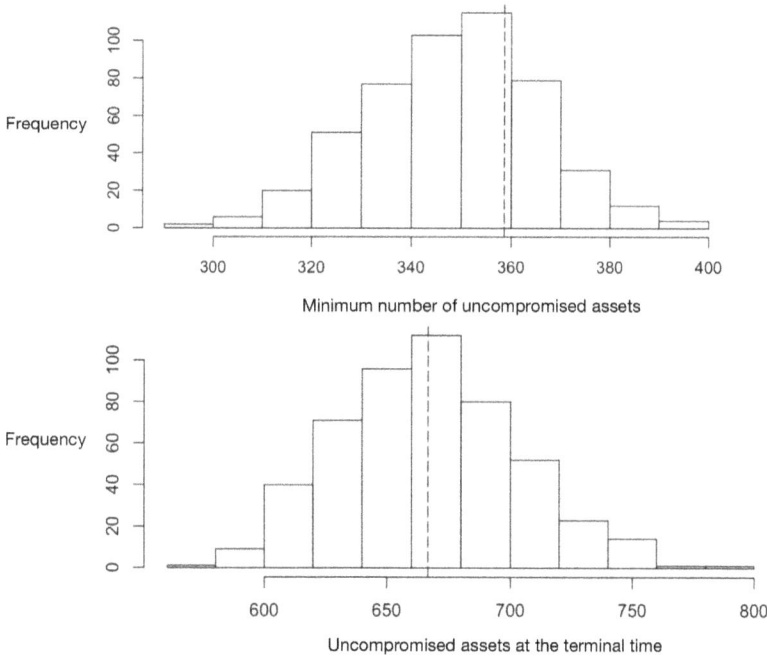

**Fig. 4.5.** We compare the solution of the ODEs and the results of the SSA for the PAM when $a_{co}$ exceeds the threshold value for the persistence of compromise by considering the distribution of the minimum number of uncompromised cyber assets (upper panel) and the number of compromised cyber assets at the time horizon (lower panel). The vertical dotted lines show the values obtained from the ODES.

long-term distribution of uncompromised cyber assets. In this case, the range is more than 200 with a peak around 670 – again a range close to 30%.

The main conclusion from these results is similar to the case in which $a_{co}$ is less than the threshold value: those managing the cyber systems that have the minimum number of compromised assets or the maximum number of uncompromised cyber assets at the terminal time are not particularly "good" but rather lucky, just the way Napoleon wanted his generals.

> **Potential project**: To address the question of whether the slow oscillations are a result of the numerical scheme that will disappear for smaller values of $dt$ or a property of the nonlinear conditional mean, code the SSA for the PAM and explore behavior for smaller values of $dt$.

## 4.4.   Stochastic Version of the FMSCO

The stochastic version of the FMSCO proceeds in a similar way as in the previous section, so I skip some of the details and all of the pseudocode.

We begin by writing Eqns. (3.1)–(3.4) in differential notation.

$$dx = (-axy + b(X_T - x - x_0))dt + o(dt) \qquad (4.14)$$

$$dx_0 = (axy - r_x x_0)dt + o(dt) \qquad (4.15)$$

$$dy = (-cxy + d(Y_T - y - y_0))dt + o(dt) \qquad (4.16)$$

$$dy_0 = (cxy - r_y y_0)dt + o(dt) \qquad (4.17)$$

Assuming that we have chosen the time interval $dt$ properly, so that the probability of only one event between $t$ and $t + dt$ is very high, there are now six transitions to consider for the random process $\{X(t), X_0(t), Y(t), Y_0(t)\}$. In analogy to Eqn. (4.8) we set $\lambda(x, x_0, y, y_0) = axy + b(X_T - x - x_0) + r_x x_0 + cxy + d(Y_T - y - y_0) + r_y y_0\}$, so that the six transitions and their probabilities are

- $\mathcal{T}_1 = \{dX = -1, dX_0 = 1, dY = 0, dY_0 = 0\}$, which occurs with probability $axy/\lambda(x, x_0, y, y_0)$;

- $T_2 = \{dX = 1, dX_0 = 0, dY = 0, dY_0 = 0\}$, which occurs with probability $b(X_T - x - x_0)/\lambda(x, x_0, y, y_0)$;
- $T_3 = \{dX = 0, dX_0 = -1, dY = 0, dY_0 = 0\}$, which occurs with probability $r_x x_0/\lambda(x, x_0, y, y_0)$;
- $T_4 = \{dX = 0, dX_0 = 0, dY = -1, dY_0 = 1\}$, which occurs with probability $cxy/\lambda(x, x_0, y, y_0)$;
- $T_5 = \{dX = 0, dX_0 = 0, dY = 1, dY_0 = 0\}$, which occurs with probability $d(Y_T - y - y_0))/\lambda(x, x_0, y, y_0)$; and
- $T_6 = \{dX = 0, dX_0 = 0, dY = 0, dY_0 = -1\}$, which occurs with probability $r_y y_0/\lambda(x, x_0, y, y_0)$.

There are now four random variables: $X_{\text{sim}}[s, n]$, $X_{0,\text{sim}}[s, n]$, $Y_{\text{sim}}[s, n]$ and $Y_{0,\text{sim}}[s, n]$. The SSA generalizes readily (which one of its powers), so I do not repeat it here. Rather, let's look at the results. For the results that follow, I used $dt = 0.0015625$.

## 4.5.   Results for the SSA Version of the FMSCO

Our first check is to compare the conditional mean dynamics, the generalization of Step 3 of the SSA, with the solution of the ordinary differential equations (Figure 4.6). These curves sit on top of each other when corrected for the offset of 1.0; the difference in the closeness of the curves in the left panels and right panels is due to the scale of the $y$-axes.

The next step, which is also a check on the choice of the time increment $dt$, is to look at individual replicates of the SSA and the solution of the ODEs (Figure 4.7). In this case, we see that the replicates surround the deterministic trajectories, and again the variation in the height of the fluctuations is due to the scale of the y-axes.

The continued analogy with the previous section is a comparison of the steady state distributions of uncompromised cyber assets, approximated by the values of the stochastic variables close to the time horizon (Figure 4.8). In addition to the distributions, I show the steady state values of the ODEs as a vertical dotted line.

In Figure 4.9, I show the correlation between uncompromised and compromised X-side and Y-side cyber assets (upper panels), uncompromised X-side and uncompromised Y-side cyber assets (lower left panel), and compromised X-side and compromised Y-side cyber

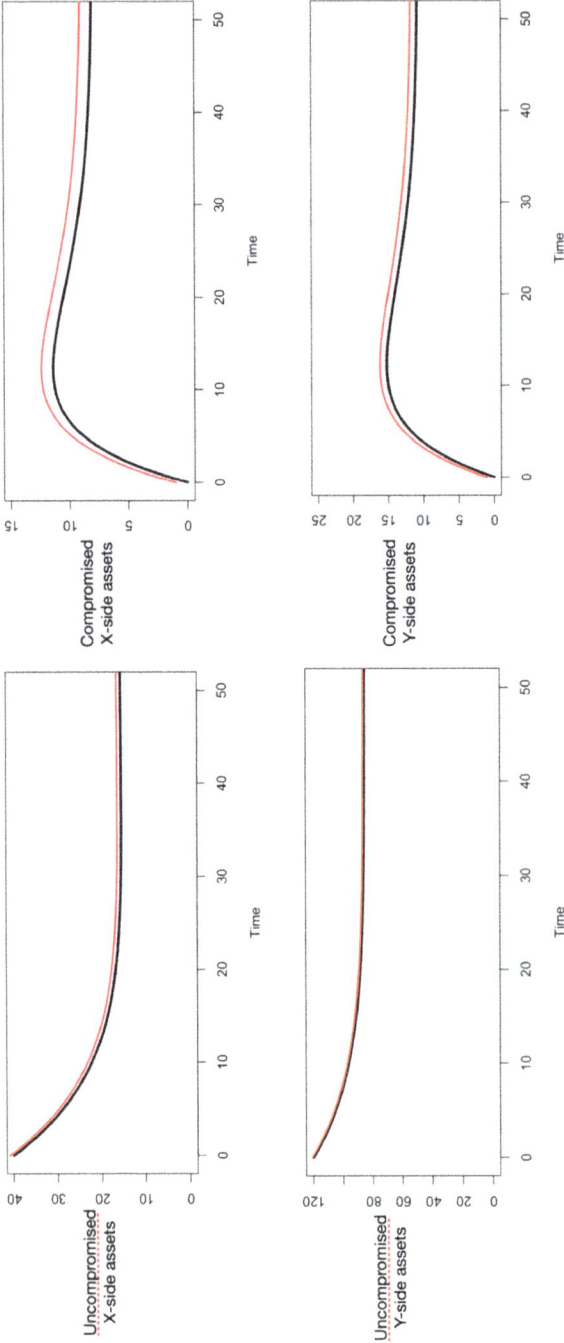

**Fig. 4.6.** As a check on the time step for the SSA version of the FMSCO, we compare the solution of the ODEs (black) and the conditional mean, generated in analogy to Step 3 of SSA (red, offset by +1.0 so that the two curves do not sit on each other for uncompromised X-side and Y-side cyber assets (left panels) and compromised X-side and Y-side cyber assets (right panels). Note the difference in the scales of the panels on the right (as well as those on the left).

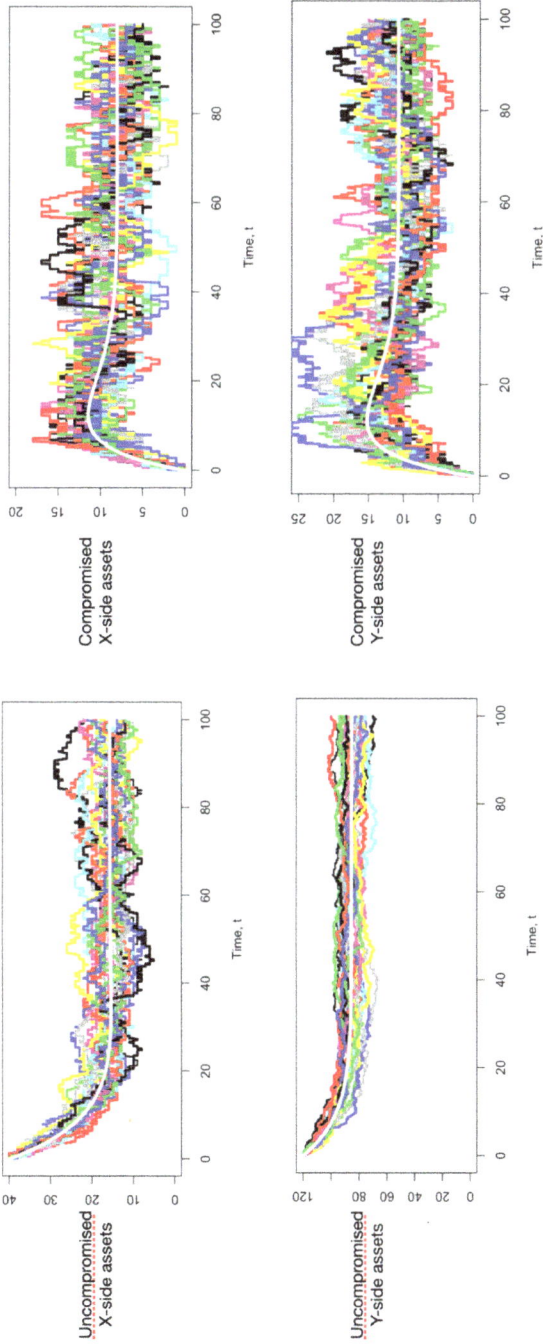

**Fig. 4.7.** The first 10 replicates of the SSA (different colors) version of the FMSCO and the solution of the ODEs (white) for uncompromised (left panels) and compromised cyber assets (right panels).

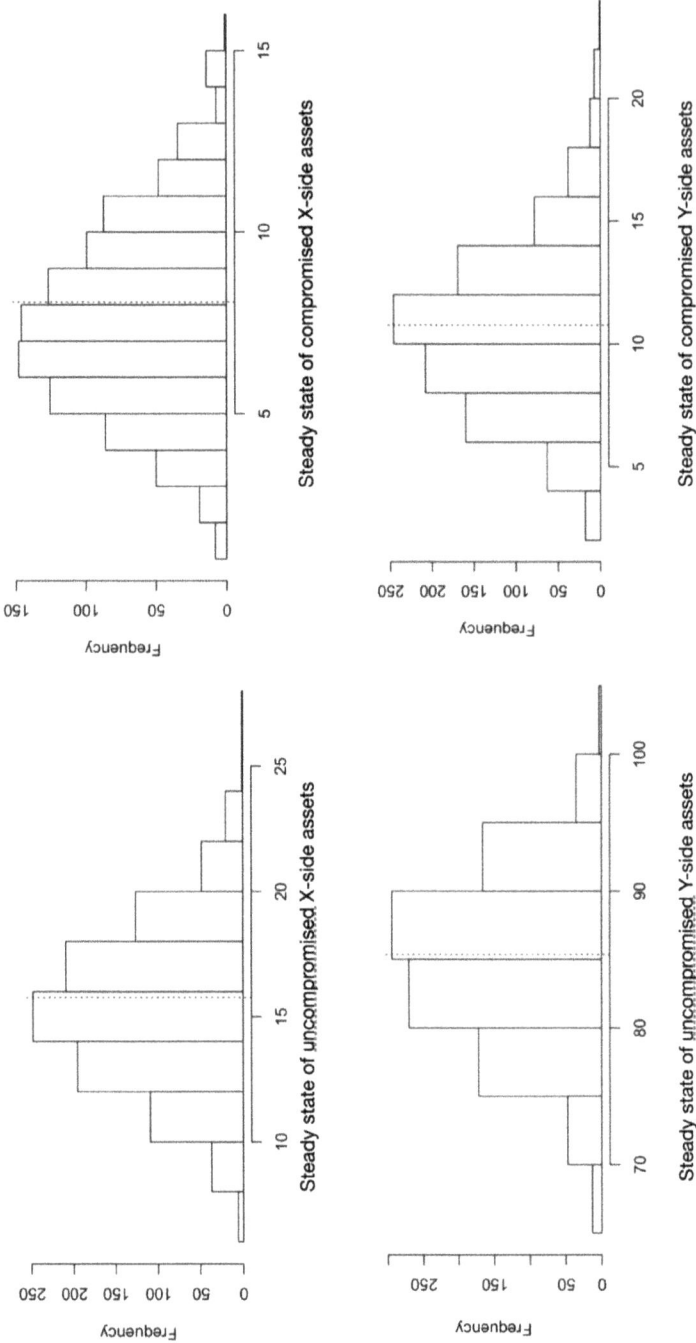

**Fig. 4.8.** Steady state distributions, determined by the values of the stochastic variables close to the time horizon, of uncompromised (left panels) and compromised (right panels) cyber assets for the SSA version of the FMSCO. As with Figures 4.3 and 4.5, the vertical dotted lines are the steady state solution of the ODEs.

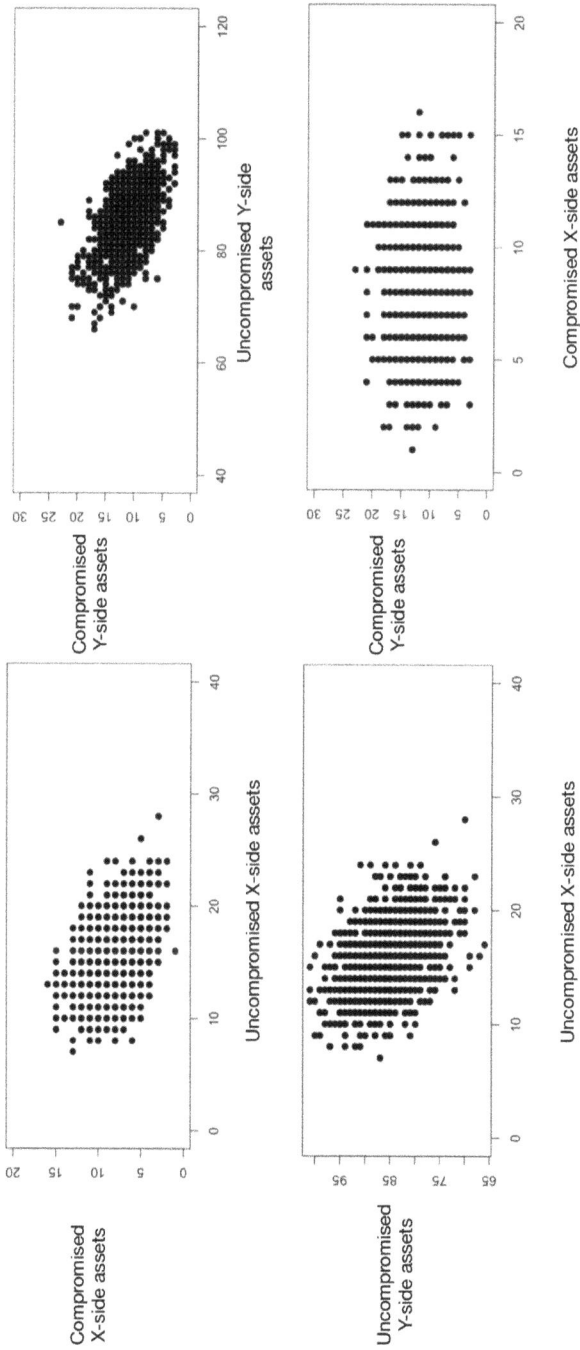

**Fig. 4.9.** The correlation between uncompromised and compromised X-side cyber assets (upper left panel), uncompromised and compromised Y-side cyber assets (upper right panel), uncompromised X-side and uncompromised Y-side cyber assets (lower left panel), and compromised X-side and compromised Y-side cyber assets (lower right panel) generated by the SSA version of the FMSCO. Since the number of cyber assets must take integer values, we see white spaces in these results. Such gaps should not be over interpreted.

assets (lower right panel) generated by the SSA version of the FMSCO. None of these results are particularly surprising, but they are comforting in confirming and quantifying our intuition.

The lower left panel in Figure 4.9 suggests an interesting insight: that the knowledge by the X-side decision maker of the number of its uncompromised cyber assets says something about the number of uncompromised cyber assets held by the Y-side. Furthermore the rate of change of the uncompromised cyber assets held by the X-side tells a decision-maker something about the rate of change of uncompromised cyber assets on the Y-side. In the lower right panel the virtually flat line through the points tells us that knowledge of the level of compromise of one's own cyber assets tells little about the level of compromise of the opponent's cyber assets (because we do not know the dynamics of the adversary's resetting pool).

> **Potential project**: If we returned to the deterministic FMSCO, the analogue of the lower left panel in Figure 4.9 would be a single line. Use Eqns. (2.2) and (2.3) to determine that line.

## 4.6.   Summary of Major Insights

- The SSA allows us to move beyond the determinism implicit in ODE models of cyber systems. In particular, the SSA allows interpretation of the ODE models in terms of the conditional mean for an underlying stochastic birth and death process.
- A stochastic version of the PAM, operationalized by the SSA, produces distributions, rather than point estimates, of quantities such as the minimum number of uncompromised cyber assets, the time to recover to nearly all uncompromised cyber assets for the case in which the rate of co-compromise is below the threshold for persistence of compromise, or the minimum number of uncompromised cyber assets and the number of uncompromised cyber assets in the steady state for the case in which the rate of co-compromise exceeds the threshold for persistence of compromise.
- A stochastic version of the FMSCO, operationalized by the SSA, produces distributions for the steady state number of compromised

and uncompromised cyber assets for each adversary and correlations between those values.

- In the stochastic version of the FMSCO knowledge by the X-side decision maker of the number of its uncompromised cyber assets says something about the number of uncompromised cyber assets held by the Y-side.
- Most importantly, for both the PAM and FMSCO using the SSA gives us a way to determine the range of variability due to fluctuations in the random processes underlying cyber attack and recovery. The SSA prevents us from misinterpreting variability as skill.

# Chapter 5

# Extensions of the Pulse Attack Model

*Learning an equation "is a journey that takes places in three stages. We begin naively without knowing the equation. We are led. . . to comprehend it, often accompanied by dissatisfaction and frustration. Finally, the experience of learning it transforms the way we experience the world"*

– Crease (2008, p. 14)

Whenever I have spoken about the population biology of disease and cyber security illustrated by the PAM and FMSCO of Chapters 2 and 3, listeners have responded enthusiastically with their ideas about how to extend the basic models. I hope that you too have extensions in mind. There are many possible extensions of the PAM and FMSCO and although I only treat some of them, this and the next chapter are a considerable fraction of the book.

In this chapter, we will consider at different levels of depth the following extensions of the PAM:

- Multiple pulse attacks over time;
- Including the perspective of the attacker, cognizant of the possibility of escalation, when choosing the attack rate parameter $a$;
- Including Cyber Protection Teams for restoring compromised assets, with associated delays in restoration;
- Allowing cyber assets to return still vulnerable to attack or temporarily hardened against attack;

- Differentiation of cyber assets according to how critical they are to performance of the cyber system or the enabled physical system;
- Attack leading to the loss of cyber assets because they are so severely damaged that they cannot be restored;
- Resources can be allocated to reducing attack, detecting compromise, or restoration;
- Adaptation of the performance function.

Each of these extensions leads to modification of either the equations for the dynamics of the cyber assets or the additional new equations. In each case, we consider motivation for the extension, modification of the model, and (in most of the cases) results.

For convenience, I repeat the equations for the PAM and performance of the cyber system or enabled physical system here:

$$\frac{dx}{dt} = -axI(t) - a_{co}xx_0 + b(X_T - x - x_0) \tag{5.1}$$

$$\frac{dx_0}{dt} = axI(t) + a_{co}xx_0 - r_xx_0 \tag{5.2}$$

$$I(t) = \frac{1}{\sqrt{2\pi}\sigma}e^{-(t_{\mathrm{peak}}-t)^2/2\sigma^2} \tag{5.3}$$

$$\phi(x) = \frac{1}{1 + e^{(x_{50}-x)/\sigma_x}} \tag{5.4}$$

Recall the behavior of these equations: when $a_{co}$ is less than the threshold value for persistence of compromise, uncompromised cyber assets decline during the attack but ultimately all cyber assets return to a fully uncompromised state. When $a_{co}$ is greater than the threshold value for the persistence of compromise, a positive fraction of the cyber assets remain compromised in the steady state. Unless otherwise noted for computations, I set $x_{50} = 400$ and $\sigma_x = 150$ in the performance function.

In a number of situations in this chapter, we have to explicitly track the number of cyber assets in the restoration pool, which we denote by $x_r(t)$. When cyber assets are neither destroyed nor added,

Eqns. (5.1) and (5.2) are equivalent to

$$\frac{dx}{dt} = -axI(t) - a_{co}xx_0 + bx_r \tag{5.5}$$

$$\frac{dx_0}{dt} = axI(t) + a_{co}xx_0 - r_xx_0 \tag{5.6}$$

$$\frac{dx_r}{dt} = r_xx_0 - bx_r \tag{5.7}$$

Summing these three equations shows that $\frac{d}{dt}(x+x_0+x_r) = 0$, which is equivalent to the assumption that the total number of cyber assets is constant, as in Eqns. (5.1) and (5.2). But when the total number of assets can change over time, we need to use Eqns. (5.5)–(5.7).

## 5.1. Multiple Pulse Attacks Over Time

The basic PAM of Chapter 2 has single pulse attack for simplicity of getting into the ideas and mathematical analysis. However, allowing multiple pulse attacks, as in Eqns. (1.2) and (1.3) and the lower panel in Figure 1.5 may sometimes provide higher fidelity to operational situations.

In this case, the model for pulse attacks becomes

$$\frac{dx}{dt} = -axI_T(t) - a_{co}xx_0 + b(X_T - x - x_0) \tag{5.8}$$

$$\frac{dx_0}{dt} = axI_T(t) + a_{co}xx_0 - r_xx_0, \tag{5.9}$$

where

$$I_T(t) = \sum_{j=1}^{J} a_j \frac{1}{\sqrt{2\pi}\sigma_j} e^{-(t_j-t)^2/2\sigma_j^2} \tag{5.10}$$

and $t_j, \sigma_j$ and $a_j$ are the time of the peak, the width, and the attack rate parameter of the $j$th pulse.

A complicating factor here is choosing the parameters of the multiple pulses. This is a situation where simulation can allow one to explore many realizations of the paths of recovery during multiple pulse attacks. It is a nice open problem for you to explore.

> **Potential project**: The simplest case of multiple attacks is the one in which there are two pulses. Even when $a_{co}$ is less than the threshold for the persistence of compromise, full recovery of the cyber system may not occur. Whether full recovery does occur or not will depend on the intensity, dispersal parameters, and timing of the two peaks. Explore recovery of the cyber system and performance as a function of these parameters, and then see if you can generalize to more than two pulse attacks.

## 5.2.   The Attacker's Perspective: Performance and the Probability of a Kinetic Response

In Chapter 3, we considered the probability that one adversary in the FMSCO will initiate a kinetic response or an attack on civilian cyber infrastructure (Eqn. (3.6)). We can do the same with the PAM by asking how the defender responds to the pulse attack and thereby explore how the attacker could choose the attack rate parameter $a$.

To do so, we sweep over values of $a$. We previously used $a = 2.0$; I will sweep over values of $a$ ranging from 0.3 to 6.0.

In Figure 5.1, I show the dynamics of the uncompromised cyber assets (left panels) and performance (right panels) for $a_{co}$ less than its threshold value (upper panels) or more than its threshold (lower panels). These results are similar to those in Chapter 2: the number of uncompromised cyber assets declines during the attack, and then either fully recovers or compromise persists according to the value of $a_{co}$. Performance, because of its sigmoidal form, may not show the same level of decline. Indeed, for small values of $a$, performance when $a_{co}$ is less than the threshold value for the persistence of compromise barely drops at all. This effect can be seen clearly if we plot the minimum value of performance as a function of $a$, as in Figure 5.2.

Why would the attacker choose a small value of $a$? One possibility is that the attacker wishes to intimidate, rather than really attack, by showing that the attacker can enter the defender's cyber system. Another possibility is limiting the upper value of $a$ is that in response to an attack that substantially degrades performance, the defender mounts a response that is either kinetic or a cyber attack on critical

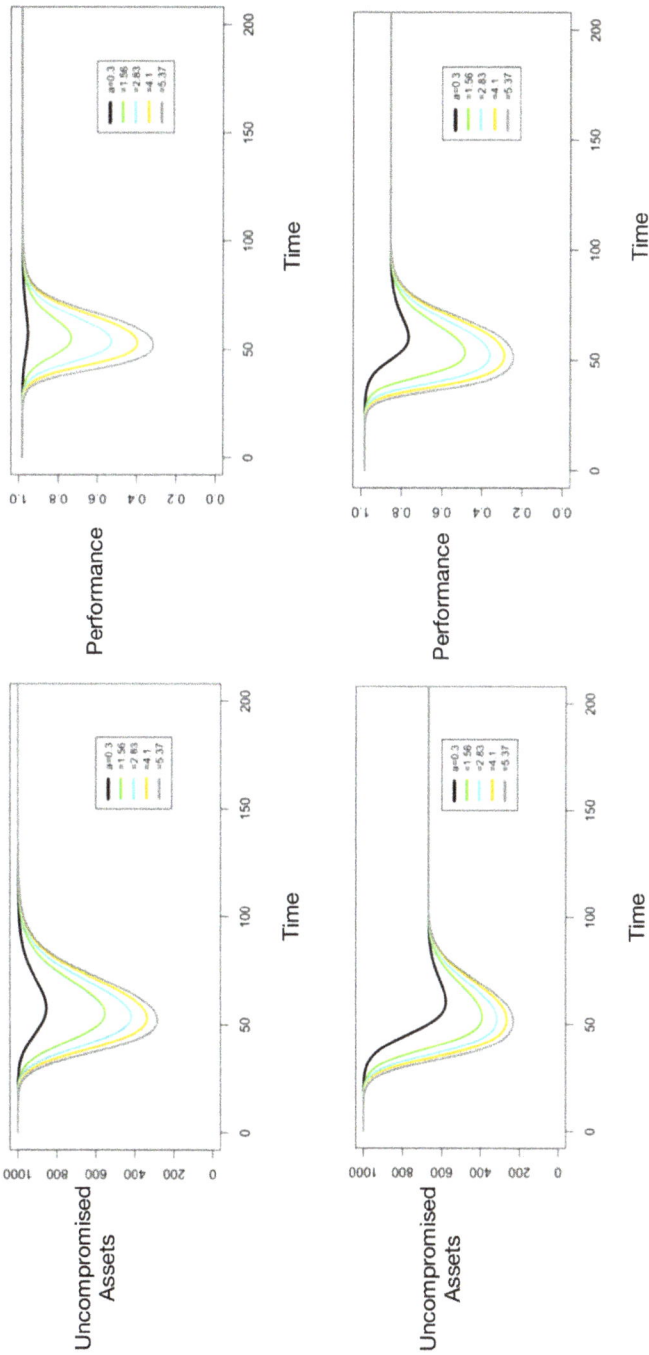

**Fig. 5.1.** The dynamics of the uncompromised cyber assets in the PAM (left panels) and performance (right panels) for $a_{co}$ less than (upper panels) or greater (lower panels) than the threshold value for the persistence of compromise.

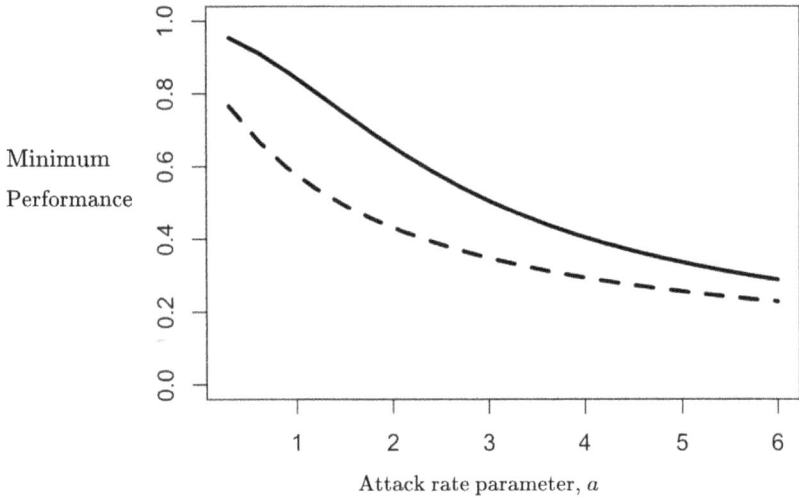

**Fig. 5.2.** Minimum performance of the cyber system or enabled physical system in the PAM as a function of the attack rate parameter $a$ when $a_{co}$ is less than the threshold (solid line) or greater than the threshold (dotted line) value for the persistence of compromise.

civilian infrastructure. We refer to either of these a response by the defender.

### 5.2.1.  *Determining the response of the defender*

A simple way of determining whether the defender will respond or not is to assume that the defender decides on action only after the peak of the pulse attack occurred. When the attacker can estimate the threshold minimum performance corresponding to a response from the defender, we can simply draw a horizontal line Figure 5.2, see where it intersects the curve, and then draw a vertical line from that intersection to the $x$-axis to determine the limit on $a$ so that the defender does not initiate a response. Determining this threshold value of minimum performance is a role for intelligence services and thus beyond the scope of this book. The assumption that the defender waits until performance has bottomed out may seem like a stretch, since it has the flavor of "Let's see how bad it gets before we take action", but is plausible for a defender who does not want to escalate the interaction.

Alternatively, once the attack begins the defender constantly monitors performance and and at each time decides whether to respond or not. Thus, we envision a function $p_r(\phi)$ giving the probability that the defender responds when performance is $\phi$. It is likely that if performance exceeds a value $\phi_1$ associated with natural variation in the cyber system (*sensu* Mangel and Brown 2022) the probability of a response will be 0 and if performance is less than a value $\phi_2$ the probability of response will be maximal, denoted by $p_{\max}$, which need not be 1. The easiest way to connect the points $(\phi_2, p_{\max})$ and $(\phi_1, 0)$ is by a straight line (Figure 5.3). We will call this the **hockey stick model** for response to cyber attack, in analogy to the shape of a hockey stick (for a recent examples of use in fisheries management see Punt *et al.* 2014, Siple *et al.* 2018, Link *et al.* 2020).

A little bit of algebra gives us the equation for the line between $(\phi_2, p_{\max})$ and $(\phi_1, 0)$

$$
p_r(\phi) = \begin{cases} p_{\max} & \phi \leq \phi_2 \\ \dfrac{p_{\max}}{\phi_1 - \phi_2}(\phi_1 - \phi) & \text{if } \phi_2 < \phi \leq \phi_1 \\ 0 & \text{if } \phi > \phi_1 \end{cases} \tag{5.11}
$$

One reason for choosing a line between $(\phi_2, p_{\max})$ and $(\phi_1, 0)$ is that there is only one line between two points, the number of curves

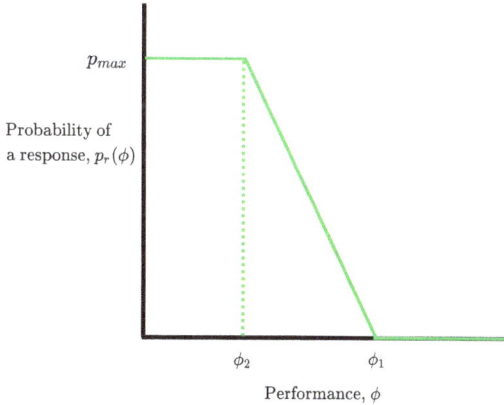

**Fig. 5.3.** The hockey stick model for response by the defender to a cyber attack (Eqn. (5.11)). In this case, the $x$-axis is performance, which ranges from 0 to 1, and the $y$-axis is the probability that the defender responds to the cyber attack.

between them is infinite. You may want to explore what happens if a curve is used instead of a line to join these points.

When the defender constantly monitors performance and decides at each time whether or not to initiate a response, we ask "What is the probability, denoted by $P_r(t)$, that the defender initiates a response at time $t$, given that a response has not be initiated yet?". Since the defender is not the aggressor, we know that $P_r(0) = 0$. To compute its value for subsequent times, we once again take advantage of the differential equations of the PAM producing solutions in discrete time. That is when $T$ is the time horizon and $dt$ is the time step, there are $N = T/dt$ values for the time variable and we let $t_n = n \cdot dt$ denote the $n$th time, where $n$ runs from 0 to $N$.

As an attack proceeds, the defender can initiate a response only once (although once it is initiated all sorts of things might happen – we put that aside for now). The probability that a response is initiated at time $t_n$ is the product of no response initiated before then, $1 - P_r(t_{n-1})$ and a response initiated at time $t_n$, which is $p_r(\phi(t_n))$. We thus obtain the iteration equation

$$P_r(t_n) = (1 - P_r(t_{n-1}))p_r(\phi(x(t_n))) \tag{5.12}$$

In Figure 5.4, I show the result of Eqn. (5.12) for the base case parameters of the PAM, when $a_{co}$ less than the threshold value for persistence of compromise. The upper left panel shows the reduction and recovery of performance during and after the pulse. The upper right panel shows the probability of a response as a function of time, determined from Eqn. (5.11) and the lower panel shows the probability of response at time $t$ given that the defender has not responded yet, determined from Eqn. (5.12). We see that $p_r(\phi(t))$ and $P_r(t)$ have similar shapes.

Using Figure 5.4 is a way of thinking along the temporal path. We can generalize the notion of thinking along the path in the following way. Each choice of the attack rate parameter $a$ corresponds to a minimum level of performance, as in Figure 5.1, and of the maximum probability of a response by the defender, as in the lower panel of Figure 5.4. We can then sweep over values of $a$, and plot maximum probability of a response by the defender versus minimum performance of the defender's cyber system or enabled physical system, as in Figure 5.5.

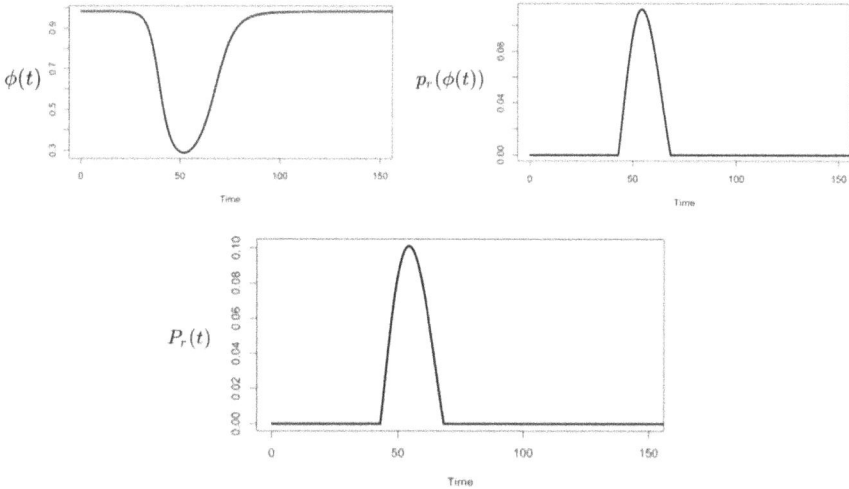

**Fig. 5.4.** Illustration of the process for computing the probability of a response to the cyber attack, Eqn. (5.12), for the base case parameters of the PAM. The upper left panel shows performance of the cyber system or enabled physical system as a function of time, with decline during the attack and recovery after the attack ends (because $a_{co}$ is less than the threshold value for the persistence of compromise). The upper right panel shows $p_r(\phi(t))$, computed from Eqn. (5.11) with $p_{max} = 0.2$, $\phi_1 = 0.85$, and $\phi_2 = 0.50$. The lower panel shows the probability that a response is initiated at time $t$, given that one has not been initiated until then, $P_r(t)$, computed from Eqn. (5.12).

This figure illustrates how the PAM can be used by the attacker in strategic considerations of the trade off between reduction in the performance of the defender's system and the probability of escalation by the defender. Although the attacker is unlikely to know the defender's response function precisely, exploring different response functions and thresholds can support assessment of the robustness of a planned attack, or assess the value of investing in intelligence to determine the defender's response function.

## 5.3. Visits by a Cyber Protection Team is Required for Restoration

In many operational situations special a **Cyber Protection Team** (CPT) needs to be called for restoring compromised assets, so that

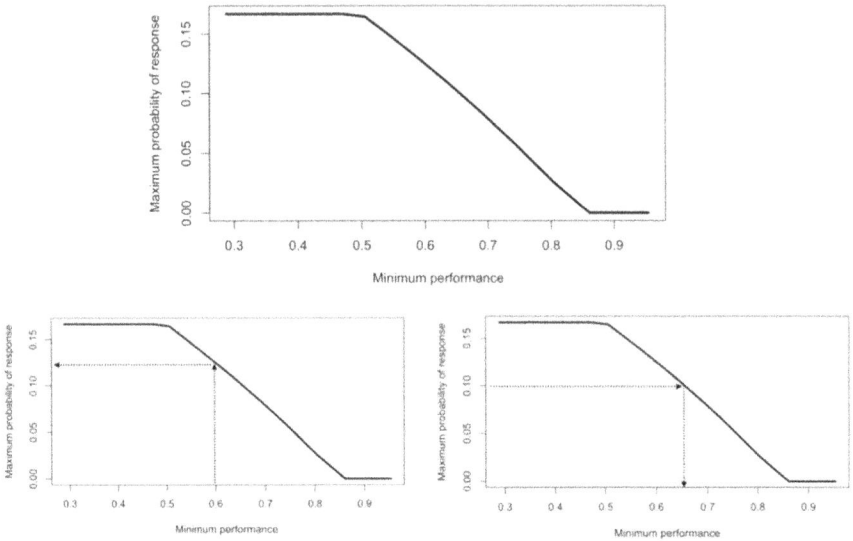

**Fig. 5.5.**   Upper panel: A sweep over values of the attack rate parameter $a$, allows us to characterize both minimum performance and the maximum probability of a response by the defender. The attacker can then use the plot in one of two ways. First, the attacker may set the target reduction in performance (lower left panel, with a target level of performance of 0.60), from there draw a vertical line until it intersects the the curve and a horizontal line from the intersection to the $y$-axis in order to determine the maximum probability of a response by the defender (here about 12.5%) Alternatively (lower right panel), the attacker may set an acceptable level for the probability of response by the defender (10% in this case), then draw a horizontal line until it intersects the curve and a vertical line from the point of intersection to the $x$-axis in order to determine the reduction in performance associated with the target level of probability of response (in this case performance drops to about 0.65).

movement from the restoring pool to the uncompromised pool is time dependent. Two ways to think about modeling CPTs are these. First, the CPTs might have regularly scheduled visits to the location (physical or virtual) of the cyber system, in which case restoration occurs only at times when the team is visiting. We can capture this idea with a function $\mathcal{C}(t)$ that is 1 when the CPT is visiting and 0 otherwise. We then replace Eqn. (5.1) by

$$\frac{dx}{dt} = -axI(t) - a_{co}xx_0 + b(X_T - x - x_0)\mathcal{C}(t) \qquad (5.13)$$

Second, CPTs may be called to visit the cyber system only when the number of cyber assets requiring restoration crosses a threshold, i.e. when $x_r = X_T - x - x_0 > x_{th}$ where $x_{th}$ is the threshold for the CPT to visit. We rearrange this condition as $X_T - x_{th} > x(t) + x_0(t)$ and let $\mathcal{H}(x, x_0|x_{th})$ denote a function that is 1 when $X_T - x_{th} > x(t) + x_0(t)$ and 0 otherwise, where the vertical bar separates the dynamical variables and the threshold for calling the CPT. The dynamics are then

$$\frac{dx}{dt} = -axI(t) - a_{co}xx_0 + b(X_T - x - x_0)\mathcal{H}(x, x_0|x_{th}) \quad (5.14)$$

You might reasonably wonder if using Eqn. (5.14) should also include a delay, since we can expect that CPTs do not instantaneously appear once the threshold is crossed. I agree, and later in the chapter we will discuss modification of the equations for the PAM when delays occur. After that, you may want to return to this section and explore some of these ideas once more.

---

**Potential project**: Develop models for the time dependent function $\mathcal{C}(t)$ in Eqn. (5.13) and the threshold function $\mathcal{H}(x, x_0|x_{th})$ in Eqn. (5.14). Then modify the basic PAM to explore the dynamics of the cyber system and performance.

---

## 5.4. Cyber Assets Restored Still Vulnerable or Temporarily Hardened to Attack

We now truly expand the PAM, in that we will move from two differential equations to three differential equations, which does not sound like a big increase until one thinks of it as a 50% increase in the complexity of the model. In particular, we assume that when cyber assets are restored a fraction $f_h$ are **hardened** against cyber attack and thus have reduced vulnerability, in that the rate at which they become compromised is $\rho aI(t)$, where $\rho < 1$. The remaining fraction $1 - f_h$ of restored assets are not hardened against attack.

We assume that (i) hardened cyber assets lose their hardening at rate $g$, (ii) the rate at which hardened cyber assets return to operational status is less than the rate $b$ at which vulnerable cyber assets are returned to operational status, and (iii) the reduction in the rate of return to operational status depends on $\rho$. In particular in the dynamics for the hardened cyber assets, we replace $b$ by $\rho^\gamma b$, to capture how hardening a cyber asset slows its return to operational status. Here $\gamma$ is a parameter that relates hardening with the rate of restoration. For example, if $\gamma = 0$, hardened cyber assets are restored at the same rate as vulnerable assets and when $\gamma = 1$, the reduction in the rate of restoration is the same as the reduction in the rate of attack. I encourage you to explore other choices of $\gamma$, thinking about the operational difference between $\gamma < 1$ and $\gamma > 1$.

Finally, we need to make a decision about the rate of co-compromise; I am going to assume that hardened and vulnerable cyber assets are co-compromised at the same rate. If you prefer different assumptions than these, that is great – and I encourage you to code up the equations and explore them.

The total number of uncompromised cyber assets, which still determines performance, is now $x(t) = x_h(t) + x_v(t)$ where $x_h(t)$ and $x_v(t)$ are the number of hardened and vulnerable uncompromised cyber assets at time $t$ respectively. The dynamics of the cyber assets are now

$$\frac{dx_h}{dt} = -a\rho x_h I(t) - a_{co}x_h x_0 - gx_h + f_h\rho^\gamma b(X_T - x_h - x_v - x_0)$$
$$(5.15)$$

$$\frac{dx_v}{dt} = -ax_v I(t) - a_{co}x_v x_0 + gx_h + (1 - f_h)b(X_T - x_h - x_v - x_0)$$
$$(5.16)$$

$$\frac{dx_0}{dt} = aI(t)[\rho x_h + x_v] + a_{co}(x_h + x_v)x_0 - r_x x_0 \qquad (5.17)$$

Assuming that hardened and vulnerable cyber assets contributed equally to performance (see the next section for an alternative), we set $x_{tot} = x_h + x_v$ so that performance is

$$\phi(x) = \frac{1}{1 + e^{(x_{50} - x_{tot})/\sigma_x}} \qquad (5.18)$$

As a check on the formulation, we set $\rho = 1$, in which case there is no difference between hardened and vulnerable cyber assets and the dynamics of uncompromised cyber assets are obtained by adding Eqns. (5.15) and (5.16)

$$\frac{d}{dt}(x_h + x_v) = -a(x_h + x_v)I(t) - a_{co}(x_h + x_v)x_0$$

$$+ b(X_T - x_h - x_v - x_0) \tag{5.19}$$

Setting $x = x_h + x_v$ we obtain Eqn. (5.1). This kind of check on your analysis is always good to do, no matter how experienced you are. When $\rho \neq 1$, the PAM is now a three-dimensional dynamical system.

For computations, I used the same parameters as in Chapter 2 for the basic PAM and appended the new parameters that capture the reduction in the rate of attack due to hardening $\rho = 0.1, 0.25, 0.50$ or 0.75, the rate at which hardening is lost $g = 0.001$, $\gamma = 1$, and the fraction of restored cyber assets that are hardened $f_h = 0.5$. For the initial conditions, I assumed no compromise at $t = 0$ and that hardened and vulnerable cyber assets were initially each $0.5X_T$.

As before, we separate the numerical results according to the value of $a_{co}$. We will focus on the dynamics of cyber assets rather than performance, to explore the effects of hardening cyber assets on system dynamics. You might want to code the equations and explore the dynamics of performance.

### 5.4.1. $a_{co}$ *is less than the threshold value for the persistence of compromise*

In this case, we expect that following the attack, cyber assets will fully restore and compromise will be extinguished. In Figure 5.6, I show the dynamics of the assets for the four values of $\rho$.

Because $a_{co}$ is less than the threshold value for persistence of compromise, after the attack ends the system recovers to a fully uncompromised state; this is captured in the gray line which sums $x_h(t)$ and $x_v(t)$. Some other noteworthy features of Figure 5.6 are:

- It is only when $\rho < 0.5$ (the two lower panels) that the decline of hardened cyber assets is noticeably slower than the decline of vulnerable cyber assets.
- Because hardening is lost continuously in time and proportional to the number of hardened cyber assets (Eqn. (5.15)) the number of

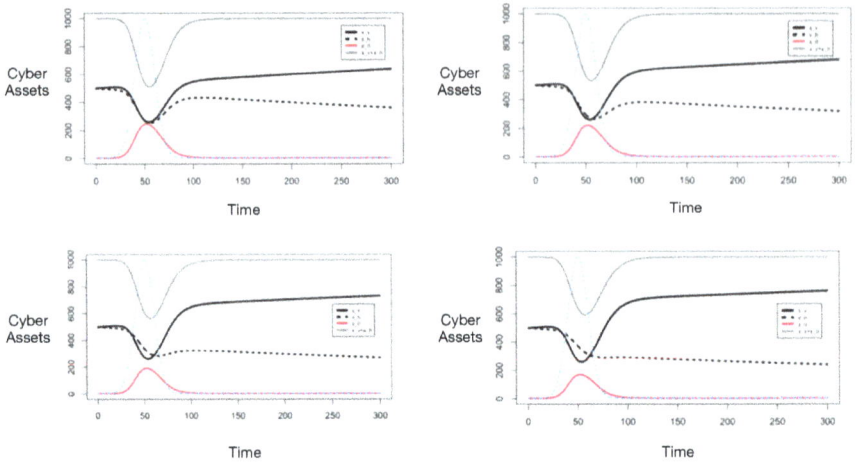

**Fig. 5.6.**    The dynamics of cyber assets for $\rho = 0.75, 0.5, 0.25$ or $0.1$ (upper left, upper right, lower left, and lower right panels, respectively) when $a_{co}$ is less than the threshold value for the persistence of compromise. The pulse attack is shown as a blue line, vulnerable uncompromised cyber assets as solid black line, hardened uncompromised cyber assets as a dashed black line, total uncompromised cyber assets as a gray line, and compromised cyber assets as a red line.

    hardened cyber assets declines slightly even before the pulse attack starts.

- For much the same reason, the number of hardened cyber assets slowly declines long after the attack.

> **Potential project**: These observations suggest modifying Eqns. (5.15) and (5.16) to include regular upgrades of vulnerable cyber assets so that they are hardened, even in the absence of attack. How would the relevant equations change? In this case, it would also be worthwhile to explore how performance depends upon the rate of upgrading when there are multiple pulse attacks.

### 5.4.2.    $a_{co}$ is greater than the threshold value for the persistence of compromise

In this case, following the attack compromise persists. In Figure 5.7, the gray lines showing the total number of uncompromised cyber

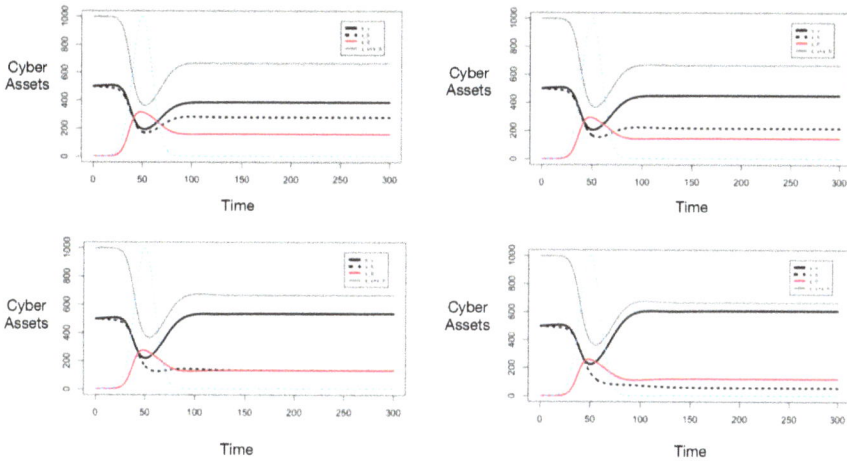

**Fig. 5.7.** The dynamics of cyber assets for $\rho = 0.75, 0.5, 0.25$, or $0.1$ (upper left, upper right, lower left, and lower right panels, respectively) when $a_{co}$ is greater than the threshold value for the persistence of compromise. The pulse attack is shown as a blue line, vulnerable uncompromised cyber assets as solid black line, hardened uncompromised cyber assets as a dashed black line, total uncompromised cyber assets as a gray line, and compromised cyber assets as a red line.

assets and the red line the number of compromised cyber assets as a function of time reach essentially similar steady state levels as $\rho$ varies.

When compared to the situation in which $a_{co}$ is less than the threshold value:

- In this case, both vulnerable and hardened cyber assets reach steady states (as opposed to the previous case in which hardened cyber assets start to decline after the pulse attack ends). Since the co-compromise rate parameter is sufficiently high to ensure that there is a continual stream of compromised cyber assets, both hardened and vulnerable cyber assets are returned from the resetting pool and thus reach a steady state.
- At first it may appear non-intuitive that smaller values of $\rho$ lead to lower steady states for the number of hardened cyber assets. I will leave this for you to explain, but offer the reminder that the rate of restoring hardened cyber assets is $\rho b$.

## 5.5.    Differentiation of Cyber Assets According to How Critical they are to Performance of the Cyber System or the Enabled Physical System

In operational settings, it is generally true that some cyber assets are more **critical to performance** of the cyber system or the enabled physical system than other assets. In the simplest case, we imagine two kinds of cyber assets, now denoted by $x_1(t)$ and $x_2(t)$, with contributions to the performance of the cyber system $v_1$ and $v_2 > v_1$, respectively. If $v_1 = 0$ those cyber assets contribute nothing to performance. Let us refer to cyber assets of type 1 as less critical or secondary, and of type 2 as critical.

Performance when there are $x_1$ and $x_2$ of each kind of cyber asset is then

$$\phi(x_1, x_2) = \frac{1}{1 + e^{\frac{x_{50} - (v_1 x_1 + v_2 x_2)}{\sigma_x}}} \qquad (5.20)$$

If we set $v_1 = v_2 = 1$ in Eqn. (5.20), we have the metric of performance we used previously.

We make these additional assumptions:

- The contributions to performance of the two cyber assets are $v_1 = 0.2$ and $v_2 = 1.0$. For computations, I assume that the total number of cyber assets of each type are $X_{1T} = 800$ and $X_{2T} = 200$.
- Cyber assets of type 1 reduce the attack rate on cyber assets of type 2. In particular, when the number of cyber assets of type 1 is $x_1$ the attack rate on cyber assets of the second kind is $e^{-\delta x_1} a I(t)$, where $\delta > 0$ is a parameter. I chose $\delta$ by assuming that when cyber assets of type 1 are at their maximum $X_{1T}$, the rate of attack on the second kind of assets is reduced to 20% of its maximum value, i.e. that $e^{-\delta X_{1T}} = 0.2$.
- There is no difference between the two kinds of cyber assets in the rates of attack or discovery of compromise, but critical cyber assets take longer to restore, in the sense that the rate of restoration of critical assets is $\rho b$, where $\rho < 1$ has a different meaning than in the previous section.
- In addition to the possibility of co-compromise, in which compromised cyber assets of type $i$ compromise cyber assets of type $i$ at rates $a_{co_1}$ and $a_{co_2}$, there may be cross co-compromise, in which

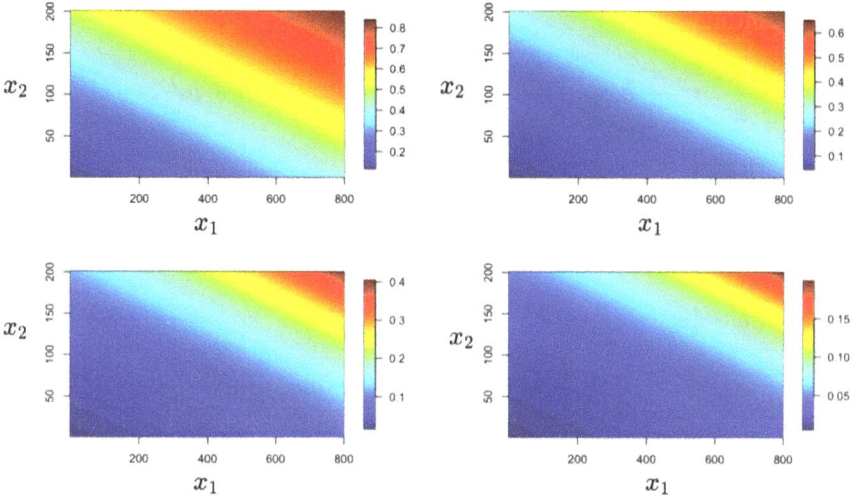

**Fig. 5.8.** When some cyber assets are more critical for performance of the cyber system or enabled physical system, performance depends upon the numbers of less critical cyber assets $x_1$ and more critical cyber assets $x_2$. Here $\sigma_x = 100$ and $x_{50} = 200$, $300$, $400$, or $500$ (upper left, upper right, lower left, and lower right panels, respectively). Note the difference in the numerical scales of the heat maps, which is not reflected in the colors.

compromised cyber assets of type $i$ compromise operational cyber assets of type $j \neq i$. We denote these rate parameters by $a_{co_{12}}$ and $a_{co_{21}}$, respectively.

In Figure 5.8, I show the performance function, Eqn. (5.20), for $\sigma_x = 100$ and four values of $x_{50}$. These heat maps are in accord with our intuition: performance is highest when the numbers of assets are closest to their maximum values, and the regions of high performance shrink as $x_{50}$ increases.

To warm up for the next section, I write the dynamics for the cyber assets using extensions of Eqns. (5.5)–(5.7). The notation extends in a natural way, so that now $x_1(t)$ and $x_2(t)$ are the numbers of uncompromised cyber assets of each type at time $t$, $x_{10}(t)$ and $x_{20}(t)$ are the numbers of compromised cyber assets of each type at time $t$, $x_{1r}(t)$ and $x_{2r}(t)$ are the numbers of cyber assets of each type in the restoring pools at time $t$, and $r_{x1}$ and $r_{x2}$ are the rates at which compromised cyber assets of each type are moved to the restoring pool. The dynamics of the two kinds of cyber assets are then

Secondary assets:

$$\frac{dx_1}{dt} = -ax_1I(t) - x_1(a_{co_1}x_{10} + a_{co_{21}}x_{20}) + bx_{1r} \qquad (5.21)$$

$$\frac{dx_{10}}{dt} = aI(t)x_1 + x_1(a_{co_1}x_{10} + a_{co_{21}}x_{20}) - r_{x1}x_{10} \qquad (5.22)$$

$$\frac{dx_{1r}}{dt} = r_{x1}x_{10} - bx_{1r} \qquad (5.23)$$

Critical assets:

$$\frac{dx_2}{dt} = -ae^{-\delta x_1}x_2I(t) - x_2(a_{co_2}x_{20} + a_{co_{12}}x_{10}) + \rho bx_{2r} \qquad (5.24)$$

$$\frac{dx_{20}}{dt} = ae^{-\delta x_1}x_2I(t) + x_2(a_{co_2}x_{20} + a_{co_{12}}x_{10}) - r_{x2}x_{20} \qquad (5.25)$$

$$\frac{dx_{2r}}{dt} = r_{x2}x_{20} - \rho bx_{2r} \qquad (5.26)$$

I will leave the check of the logic to you, that is to determine the conditions in which these equations collapse to Eqns. (5.1) and (5.2).

When solving Eqns. (5.21)–(5.26), I set the cross co-compromise rate parameters equal to 0, i.e. $a_{co_{21}} = a_{co_{12}} = 0$. I then tried to match the other parameters with those in Chapter 2, i.e. $X_{1T} = 800, X_{2T} = 200$ (so that there the total number of cyber assets is 1000), $a = 0.2, b = 0.2$ and $r_{x1} = r_{x2} = 0.2$ so that the rates of attack, restoration of cyber assets of type 1, and movement of compromised cyber assets to the restoration pool are the same as before. I set $\rho = 0.5$ and $\delta = 0.002$. To determine the co-compromise rate parameters, I set thresholds for co-compromise to be $a_{co_{1thr}} = r_{x1}/X_{1T}$ and $a_{co_{2thr}} = r_{x2}/X_{2T}$, which would be the threshold levels of the co-compromise rate parameters if we only had cyber assets of type 1 or type 2. I continue to refer to these as the threshold values for the persistence of compromise and set co-compromise rate parameters to be 0.5 or 1.5 times the values of the thresholds. We will only explore consequences when the co-compromise rate parameters are either both below or both above these threshold values; I leave the other cases to you.

The newest concepts in Eqns. (5.21)–(5.26) are $\delta$, the protection of critical cyber assets provided by secondary cyber assets, and $\rho$, the reduction (relative to secondary cyber assets) in the rate of restoration of critical cyber assets. The best situation for the cyber system is that both $\delta$ and $\rho$ are high – the secondary cyber assets provide more protection for larger values of $\delta$ (Eqn. (5.24)) and critical cyber assets are restored to operational status with larger values of $\rho$ (although we require $\rho \leq 1$). Note that the steady states of Eqns. (5.21)–(5.23) are independent of both $\delta$ and $\rho$, and the steady states of Eqns. (5.24)–(5.26) are independent of $\delta$. We will focus on minimum performance during the pulse attack.

We will sweep over values of $\delta$ and $\rho$ and examine performance of the cyber system or enabled physical system. I let $\rho$ range from 0.1 (critical cyber assets require roughly an order of magnitude longer to be restored than secondary assets) to 1.0 (critical cyber assets are restored at the same rate as secondary cyber assets) and $\delta$ range from $6.412 \cdot 10^{-5}$ to $3.745 \cdot 10^{-3}$ so that the rate of attack when secondary cyber assets were at their maximum ranged between 5% and 95% of the rate of attack if there were no protection provided by the secondary cyber assets.

### 5.5.1. *Dynamics of cyber assets, performance, and tradeoffs when the co-compromise rate parameters are less than the threshold values for the persistence of compromise and there is no cross co-compromise*

In Figure 5.9, I show the dynamics of cyber assets (upper panels) and performance (lower panels) for the four values of $x_{50}$ in Figure 5.8, both as absolute values (left panels) and relative values (right panels).

In the upper panels, the solid lines correspond to secondary cyber assets and the dotted lines to critical cyber assets; black lines are uncompromised cyber assets and red lines are compromised assets. The dotted blue line is the pulse. The role of secondary cyber assets in reducing the rate of attack on critical cyber assets is made clear by considering the relative numbers of each kind of cyber asset, i.e. $x_1(t)/X_{1T}$ and $x_2(t)/X_{2T}$, as in the upper right panel. The decline of critical cyber assets lags that of secondary assets.

The pattern of decline and recovery of performance in the lower left panel, where the pulse is a faint gray line, as a function of

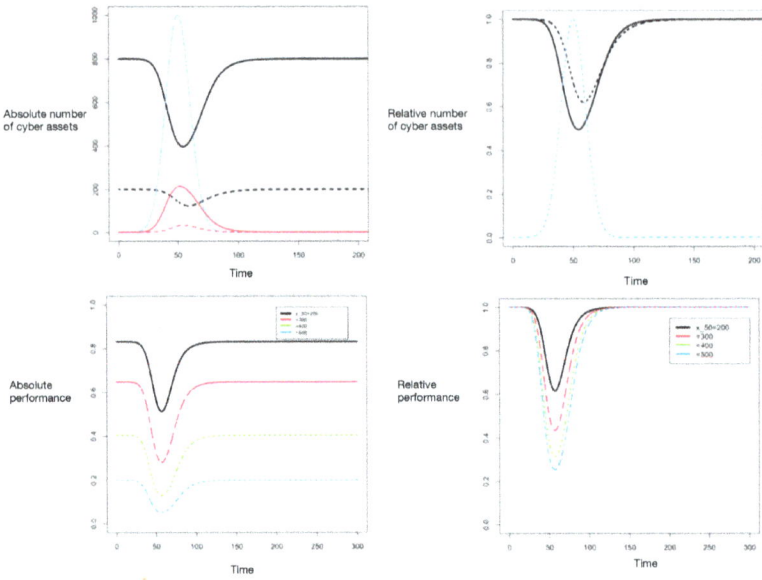

**Fig. 5.9.** The dynamics of the numbers of cyber assets (upper left panel) and performance (lower left panel) and their relative values (right panels) when the co-compromise rate parameters are less than the thresholds for the persistence of compromise. In the upper panels, the solid lines correspond to secondary cyber assets and the dotted lines to critical cyber assets; black lines are uncompromised cyber assets and red lines are compromised cyber assets. The dotted blue line is the pulse. The role of secondary cyber assets in reducing the rate of attack on critical cyber assets is made clear by considering the relative numbers of each kind of cyber asset, i.e. $x_1(t)/X_{1T}$ and $x_2(t)/X_{2T}$, as in the upper right panel. The decline of critical cyber assets lags that of secondary assets. The pattern of decline and recovery of performance in the lower left panel, where the pulse is a faint gray line, as a function of $x_{50}$ is consistent with Figure 5.8. When we consider relative performance (lower right panel), we see that there is an interaction between the value of $x_{50}$ and the minimum value of relative performance, which could not be anticipated.

$x_{50}$ is consistent with our intuition from Figure 5.8. When we consider relative performance (lower right panel), we see that there is an interaction between the value of $x_{50}$ and the minimum value of relative performance, which could not be anticipated (or maybe it could – what is your intuition?).

In Figure 5.10, I show minimum performance of the cyber system or enabled physical system for the four values of $x_{50}$ in the performance function. Some aspects of the figure are easily explained.

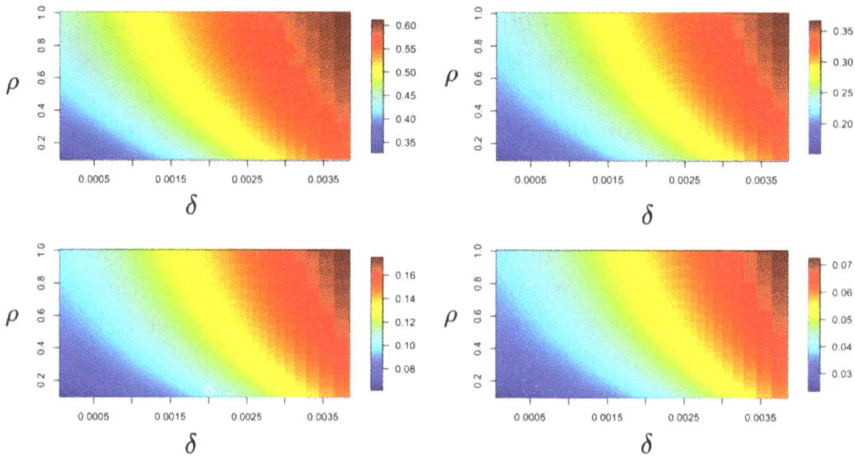

**Fig. 5.10.** Minimum performance of the cyber system or enabled physical system for the co-compromise rate parameters less than the thresholds for the persistence of compromise, as we sweep over values of $\delta$ and $\rho$ in Eqns. (5.21)–(5.26) for $\sigma_x = 100$ and $x_{50} = 200, 300, 400$, or $500$ (upper left, right, lower left, and lower right panels, respectively) in the performance function in Eqn. (5.20). Note the difference in the scales.

For example, the lower left corner – where $\delta$ and $\rho$ are the smallest – shows the worst performance, and the upper right corner shows the best performance. The four panels are qualitatively similar (although there is a shrinking of the best performance region as $x_{50}$ increases, but the quantitative metrics (i.e. the scale of the heat map) differ considerably – by almost an order of magnitude as $x_{50}$ increases from 200 to 500. This is not surprising in light of the performance function Eqn. (5.20), but it does raise an interesting question: How would this figure change if we scaled minimum performance by initial performance, $\phi(X_{1T}, X_{2T})$? I leave this question for you to answer.

> **Potential project**: As we discussed in Chapter 2, even if there is a full recovery of the cyber system and performance following the pulse attack, the rate at which recovery occurs – resilience – may vary with parameters. How does resilience vary with $\delta$ and $\rho$? Before computation, think about the problem and make qualitative predictions.

### 5.5.2.    *Dynamics of cyber assets, performance, and tradeoffs when the co-compromise rate parameters are greater than the threshold values for the persistence of compromise and there is no cross co-compromise*

Figure 5.11 is the analogue of Figure 5.9. As expected, neither cyber assets nor performance fully recover following the end of the pulse attack. Somewhat surprising (to me at least) is that critical cyber assets and performance show a small transient during recovery (best seen in the right panels).

Figure 5.12 is the analogue of Figure 5.10. The qualitative and quantitative patterns are similar to when the co-compromise rate parameters are smaller than the threshold values for the persistence of compromise. The quantitative values are smaller than in the previous case because once the attack starts the number of uncompromised cyber assets is reduced by both external attack and internal co-compromise, with the associated effect on performance.

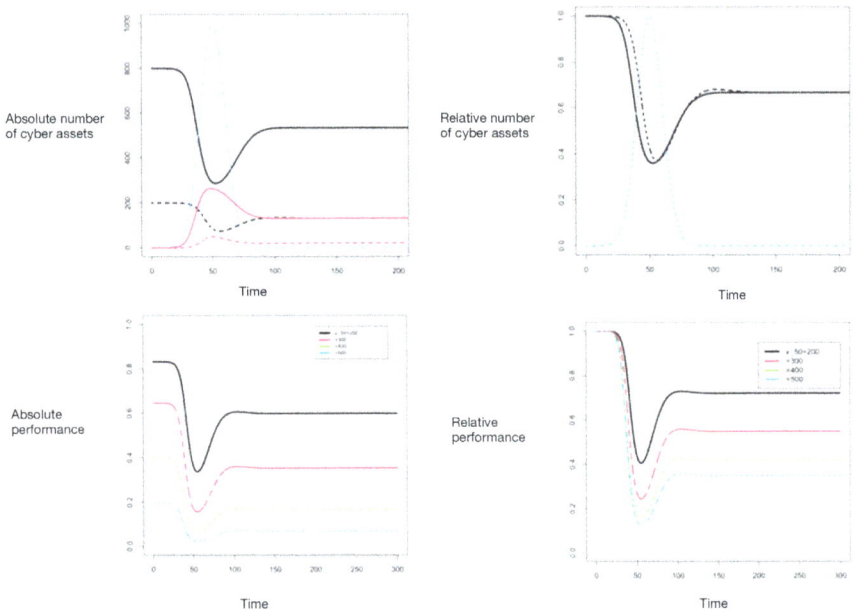

**Fig. 5.11.**    The analogue of Figure 5.9 when the co-compromise rate parameters are greater than the thresholds for the persistence of compromise.

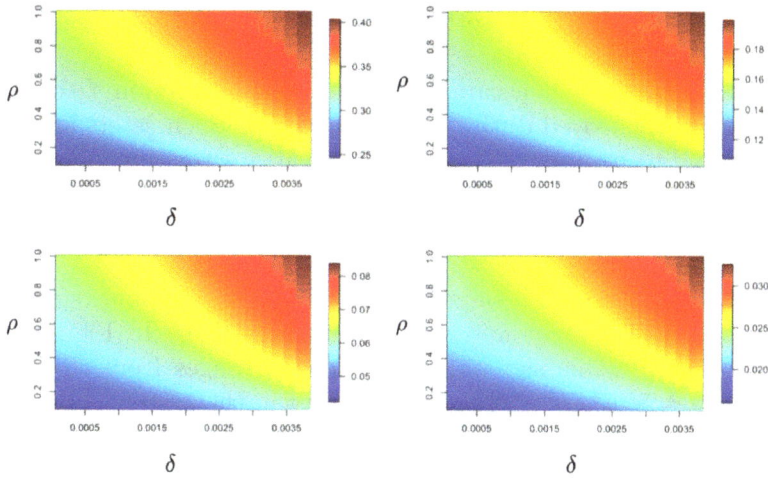

**Fig. 5.12.** The analogue of Figure 5.10 when the co-compromise rate parameters are greater than the thresholds for the persistence of compromise.

These figures raise the question: in a world of limited resources, how do we pick the "best" (or at least good) values of $\delta$ and $\rho$? We will explore a simpler version of this of this question subsequently in this chapter. Before that, however, we explore what happens when some cyber assets may be destroyed during the pulse attack.

> **Potential project**: Even without cross co-compromise, the parameter space could be expanded simply by choosing one of $a_{co_1}$ or $a_{co_2}$ to be larger than the threshold for the persistence of compromise and the other less than the threshold for the persistence of compromise. When we allow for cross co-compromise, we need to choose $a_{co_{21}}$ and $a_{co_{12}}$ as well. Intuition suggests that a good starting point has them less than the co-compromise rate parameters, but there is still a lot of work to be done here!

## 5.6. Cyber Assets may be Permanently Destroyed During Attack

Until now, we assumed that cyber assets are neither added nor removed from the cyber system, which allowed us to write the number of assets in the resetting/restoring pool at time $t$ as $x_r(t) = X_T - x(t) - x_0(t)$.

We now relax that assumption, and allow a fraction $f_d$ of cyber assets to be destroyed during the pulse attack. Such assets are colloquially said to be "bricked", because the attack turns a computer or other cyber device into nothing more than a brick. Thus, the total number of cyber assets varies over time, and we have to treat the size of the restoring pool explicitly, as in Eqns. (5.5)–(5.7).

We continue to use a sigmoidal performance function, but in this case keep $x_{50}$ constant and vary $\sigma_x$ in Eqn. (5.4). In Figure 5.13, I show four possible sigmoids, in which $\sigma_x$ ranges from 50 to 300. When $x_{50} = 50$ there is great redundancy in the system – the number of uncompromised cyber assets can fall to less than 600 before performance begins to drop. On the other hand, when $x_{50} = 300$ performance smoothly declines (in a nearly linear fashion). Note that if $\sigma_x$ is sufficiently large the performance of the cyber system or the enabled physical system is positive even when there are no uncompromised cyber assets. This is more likely to occur with a physical system that has backup technology than an only cyber system with a physical backup (e.g. a cyber communications system that also has a stand alone radio as a backup). We can envision measuring the performance function by conducting operational experiments and there are reasons to expect a range of values of $\sigma_x$, which is why we will consider the range of values here.

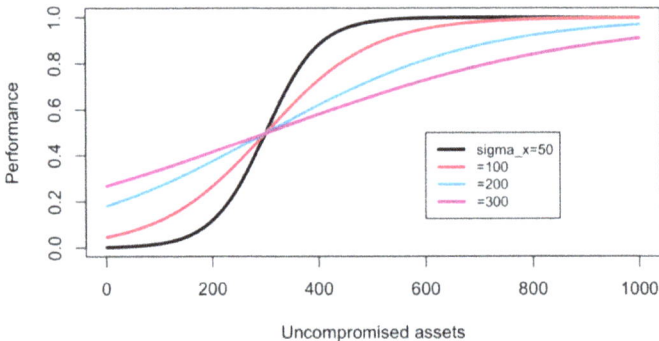

**Fig. 5.13.** Four sigmoidal performance functions with the same value $x_{50} = 300$ but different values of the dispersion parameter $\sigma_x$ (denoted by $\sigma_x$ in the caption).

When cyber assets are destroyed during an attack, even if $a_{co}$ is less than the threshold for co-compromise, the number of uncompromised cyber assets and performance will no longer return to their values at the start of the attack because the number of remaining uncompromised cyber assets will be smaller than the original number.

### 5.6.1. *The disease analogy*

The analogy of cyber assets being destroyed during an attack is like individuals dying in the SIR or SIRS models. In disease modeling, we often also include reproduction and mortality from the disease. For example, if only susceptible individuals reproduce and only infected individuals die, the SIRS equations become

$$\frac{dS}{dt} = rS(t) - \beta I(t)S(t) + \gamma R(t)$$

$$\frac{dI}{dt} = \beta I(t)S(t) - (\mu + m)I(t)$$

$$\frac{dR}{dt} = \mu I(t) - \gamma R(t)$$

The new terms on the right sides of these equations are $rS(t)$ in the equation for the rate of change of susceptible individuals and $mI(t)$ for infected individuals. Both parameters $r$ and $m$ are positive: $r$ is the per-capita (per-individual) birth rate in the population and $m$ is the per-capita rate of mortality from the disease. Since $\mu + m \geq \mu$ more individuals leave the infected pool than move to the recovery pool.

How would you modify these equations if individuals – regardless of their disease status – were subject to additional, non disease-related mortality? If individuals, regardless of disease status reproduced?

### 5.6.2. *The consequences of cyber assets being destroyed during the pulse attack*

When a fraction $f_d$ of cyber assets are destroyed during the pulse attack, we replace Eqn. (5.6) by $\frac{dx_0}{dt} = ax(1 - f_d)I(t) + a_{co}xx_0 - r_x x_0$. We now have to decide what happens when cyber assets are destroyed. Three possibilities of increasing complexity are

- Destroyed cyber assets are not replaced, so that the total number of cyber assets declines in time. In this case, we are interested in the interaction between $f_d$ and $\sigma_x$.
- Destroyed cyber assets are replaced from an at-hand pool of uncompromised assets, whose size at time $t$ is denoted by $x_p(t)$, with initial size $X_P$. Since the replacement cyber assets are at-hand, we assume that replacement of destroyed cyber assets is virtually instantaneous. In this case, there will be interaction between $f_d$, $X_P$, and $\sigma_x$.
- Destroyed cyber assets are replaced from a pool of uncompromised assets that is physically remote from the focal cyber system so that there is a time delay $\tau$ in replacing the destroyed cyber assets. We will assume that the remote pool is sufficiently large that we can ignore the possibility of its exhaustion. Now the interaction of interest is between $f_d$, $\tau$, and $\sigma_x$. Furthermore, we now have to deal with an entirely new kind of model – a differential equation with a delay.

For computations, I set $a_{co}$ less than the threshold for persistence of compromise, because we know that in this case if no assets were destroyed during the cyber attack, cyber assets and performance will return to their values before the attack started. For illustrative dynamics, I use $f_d = 0.1$, and for sweeps over the fraction of cyber assets destroyed let $f_d$ range from 0.01 to 0.3 in increments of 0.01.

> **Possible project**: Repeat the calculations we are about to do for the situation in which $a_{co}$ exceeds the threshold for the persistence of compromise.

### 5.6.3.   *Destroyed cyber assets are not replaced*

When destroyed cyber assets are not replaced, we replace Eqns. (5.5)–(5.7) by

$$\frac{dx}{dt} = -axI(t) - a_{co}xx_0 + bx_r \tag{5.27}$$

$$\frac{dx_0}{dt} = ax(1 - f_d)I(t) + a_{co}xx_0 - r_x x_0 \tag{5.28}$$

$$\frac{dx_r}{dt} = r_x x_0 - bx_r \tag{5.29}$$

In Figure 5.14, I show the dynamics of cyber assets and performance. We track uncompromised cyber assets $x(t)$, compromised cyber assets $x_0(t)$, restoring cyber assets $x_r(t)$ and their total $x_{tot}$, which will be less than $X_T$ because of the destroyed cyber assets. The steady state number of cyber assets (all of which are uncompromised) is about 880 for $f_d = 0.1$. This suggests a rule of thumb such

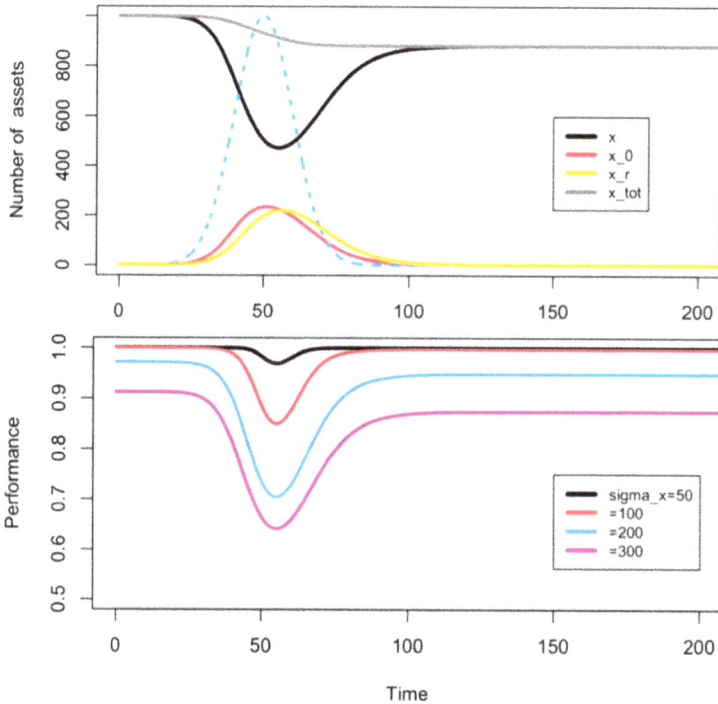

**Fig. 5.14.** Dynamics of cyber assets (upper panel) and performance (lower panel) during a pulse attack with $a_{co}$ less than its threshold value for the persistence of compromise and $f_d = 0.1$, so that 10% of the compromised cyber assets are destroyed as a result of attack. We now track uncompromised $x(t)$, compromised $x_0(t)$, and restoring cyber assets $x_r(t)$ and their total $x_{tot}$, which will be less than $X_T$ because of the destroyed cyber assets.

as "after the attack the fraction of remaining assets is about $1 - f_d$". We can explore this rule of thumb by sweeping over values of $f_d$.

Regarding performance (the lower panel of Figure 5.14): $\sigma_x = 50$ or 100 leads to a small or modest reduction in performance during the pulse attack but a full or nearly full recovery of performance (even though there is a reduction in the total number of cyber assets) long after the attack ends, while $\sigma_x = 200$ or 300 leads ultimately to recovery of uncompromised cyber assets but at a lower number than their initial number, so that performance does not fully recover (especially for $\sigma_x = 300$).

A sweep over $f_d$ shows us that both minimum performance (Figure 5.15, upper panel) and steady state performance (Figure 5.15, lower panel) decline as $f_d$ increases. For both minimum

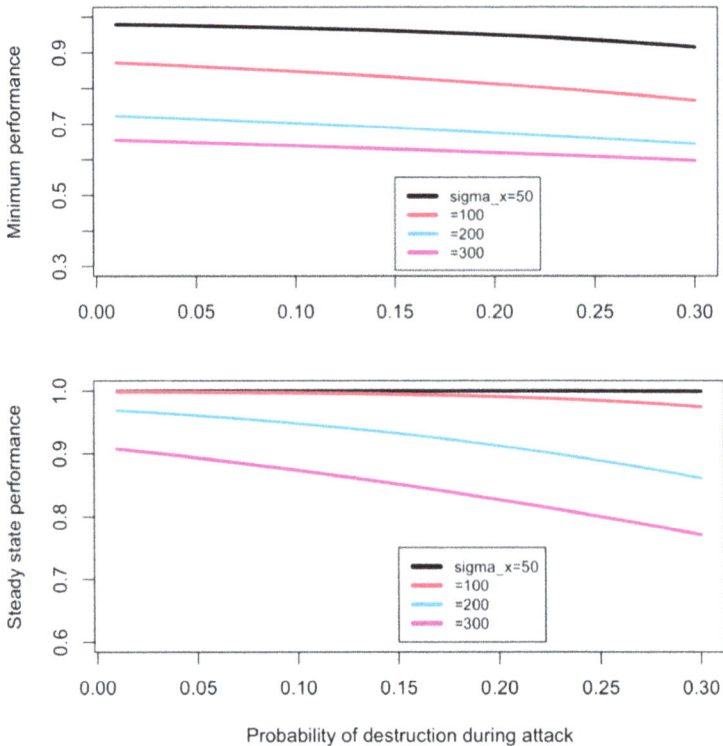

**Fig. 5.15.** Minimum (upper panel) and steady state performance (lower panel) as the probability that a cyber asset is destroyed in the attack $f_d$ ranges from 0.01 to 0.3.

and steady state performance, we see slight nonlinearities in the decline (more so for steady state than minimum performance). Also note that as $f_d$ increases, there is separation of recovery in the steady state, in that for large values of $f_d$, recovery of performance is still complete for $\sigma_x = 50$ but is not for $\sigma_x = 100$.

> **Potential project**: Code the relevant extension of the PAM and sweep over values of $\sigma_x$, for example from $\sigma_x = 50$ to 400 in steps of 25 or 50, and then make heat maps of minimum and steady state performance with $f_d$ and $\sigma_x$.

### 5.6.4. *Destroyed cyber assets are replaced instantaneously from an on-hand reserve pool*

When destroyed cyber assets are replaced from an on-hand pool of reserve uncompromised cyber assets, we append the contribution of the replacement pool to Eqn. (5.27), taking into account that the pool can be exhausted (i.e. it is possible to have a replacement pool with 0 cyber assets in it) so that we have

$$\frac{dx}{dt} = -axI(t) - a_{co}xx_0 + bx_r + \min(x_p, axf_dI(t)) \tag{5.30}$$

$$\frac{dx_0}{dt} = ax(1 - f_d)I(t) + a_{co}xx_0 - r_xx_0 \tag{5.31}$$

$$\frac{dx_r}{dt} = r_xx_0 - bx_r \tag{5.32}$$

$$\frac{dx_p}{dt} = -axf_dI(t), \text{ with the requirement that } x_p(t) \geq 0 \tag{5.33}$$

In order to choose the initial size for the pool, $X_P$, let us consider the dynamics of cyber assets that are destroyed during the attack. Denoting them by $x_d(t)$, we have $x_d(0) = 0$ because no cyber assets are destroyed before the pulse attack starts and

$$\frac{dx_d}{dt} = axf_dI(t) \tag{5.34}$$

which has the solution $x_d(t) = f_d \int_0^t ax(s)I(s)ds$. We again treat the integral as the sum of small increments. Using the same notation as previously, we let $t_n = n \cdot dt$, where $n$ runs from 0 to $N = T/dt$, and we can write simple iteration equation for $x_d(t)$. Namely, $x_d(0) = 0$ and $x_d(t_n) = x_d(t_{n-1}) + ax(t_n)f_dI(t_n)dt$ (which is a discrete form of the differential equation Eqn. (5.34)).

When we solve Eqn. (5.34) for a single value of $f_d$, we obtain the trajectory of the cyber assets destroyed during the pulse attack (upper panel of Figure 5.16); if we sweep over values of $f_d$, the values of $x_d(T)$ are the cumulative number of destroyed cyber assets (lower panel of Figure 5.16). In the lower panel, there is a nearly linear

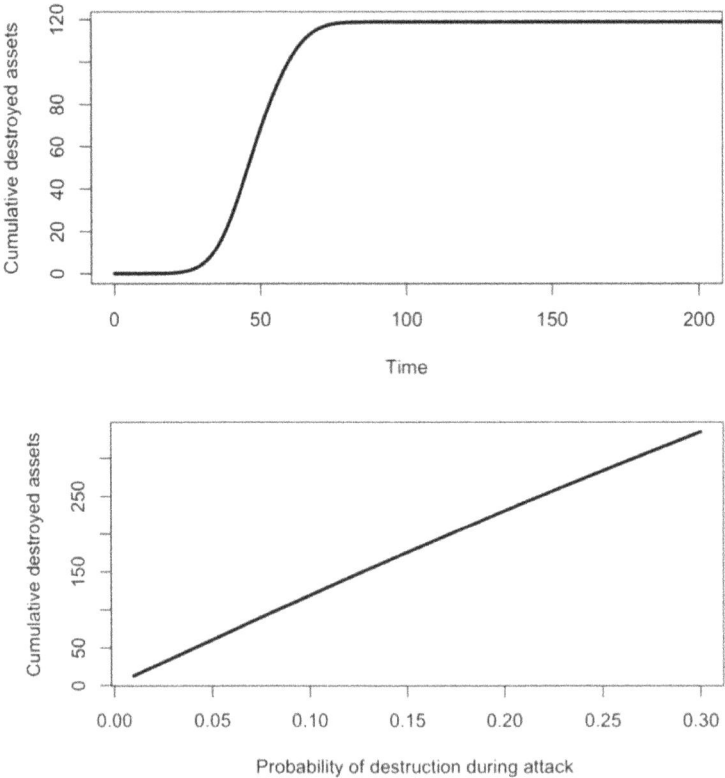

**Fig. 5.16.**  Upper panel: Choosing a single value of $f_d$ (here 0.1) allows us to follow the trajectory of uncompromised cyber assets that are destroyed during the attack. Lower panel: Sweeping over values of $f_d$ and plotting $x_d(T)$ as a function of $f_d$ allows us to see the wider relationship between the probability of destruction of an uncompromised cyber asset and the cumulative number of cyber assets destroyed during the attack.

relationship between $x_d(T)$ and $f_d$ over the entire range of $f_d$; we could expect that because Eqn. (5.34) is linear in $f_d$. Fitting a line that goes through the origin (when $f_d = 0$ no assets are destroyed during the attack) the upper point in the lower panel, gives $x_d(T) = 1150 f_d$. These results suggest that $X_P$ around 900–1200 times $f_d$ will likely give further insights into the consequences of destruction of cyber assets. To begin, we consider the dynamics of the cyber assets (Figure 5.17). The most relevant trajectory is the one corresponding to the pool ($x_p(t)$, shown in gold). Here, we see that when the initial pool size is too small, the pool exhausts before the attack has ended ($X_p = 180, 200$, or $220$) and that when the pool is sufficiently large ($X_p = 240$) the pool stays positive, but is almost exhausted, during the attack. Very large values of $X_p$ lead to cyber assets remaining in the pool after the attack ends. In this figure, we also see that the steady state number of uncompromised cyber assets is less than $X_T$ when the pool exhausts but $X_T$ when the pool does not exhaust.

In Figure 5.18, I show performance when $f_d = 0.2$, the dispersal parameter varies from 50 to 300 in increments of about 2.5, and the size of the replacement pool varies $X_P$ varies from 150 to 250. I also show the contours for which performance is 0.95, 0.85, and 0.75. When $X_P$ is sufficiently large, the contours are essentially vertical lines, which we understand to mean that with sufficiently

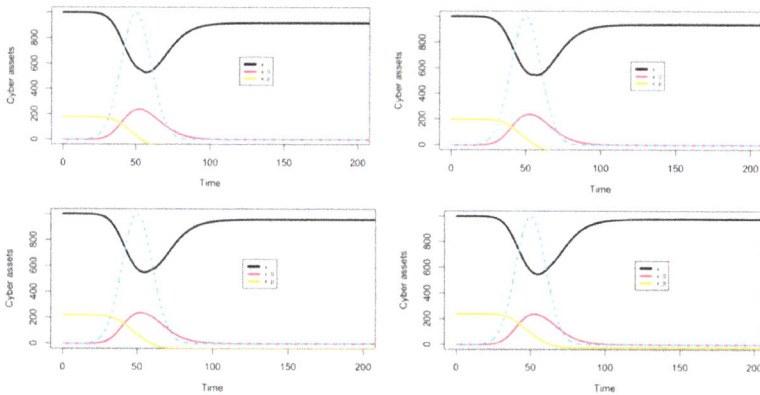

**Fig. 5.17.** The dynamics of uncompromised (black), compromised (red), and cyber assets in the pool of uncompromised replacement cyber assets (gold) for pool sizes of 180, 200, 220, and 240 cyber assets (upper left, right; lower left, right, respectively) when $f_d = 0.2$.

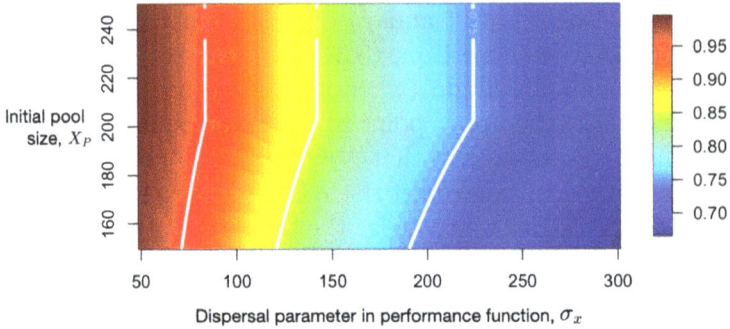

**Fig. 5.18.**    Steady state performance when $f_d = 0.2$, the dispersal parameter in the performance function varies from 50 to 300, and initial size of the replacement pool varies from 150 to 250 cyber assets. The contours for which performance is 0.95, 0.85, and 0.75 are the white lines, moving from left to right.

large reserve pools performance will only depend upon the dispersal parameter in the performance function. As $X_P$ becomes smaller – corresponding to possible exhaustion of the replacement pool – the contours of performance bend to the left, which we understand to mean a more knife edge performance function is needed to maintain the same level of performance by keeping the exponents in Eqn. (5.4) positive.

This figure begs the question: is it possible to make the dispersal parameter smaller, and thus performance larger for the same number of cyber assets; we will return to this question in the last section of the chapter.

> **Potential project**: Conduct a sweep over $f_d$ and $X_P$ and examine how the minimum and steady state numbers of uncompromised cyber assets, and minimum and steady state performance depend upon $f_d$ and $X_P$. Then think about how you would incorporate a cost of maintaining the pool.

### 5.6.5.    *Destroyed cyber assets are replaced from a storage facility, leading to a delay*

We now assume that there is a lag $\tau$ between the time that a cyber asset is destroyed and replaced so that at time $t$ uncompromised cyber assets are increased at rate $bx_r(t)$ from the restoring pool and at rate $ax(t - \tau)f_d I(t - \tau)$ from the replacement pool. Hence the dynamics of cyber assets are

$$\frac{dx}{dt} = - ax(t)I(t) - a_{co}x(t)x_0(t) + bx_r(t) + ax(t - \tau)f_d I(t - \tau)$$
(5.35)

$$\frac{dx_0}{dt} = ax(t)(1 - f_d)I(t) + a_{co}x(t)x_0(t) - r_x x_0(t)$$
(5.36)

$$\frac{dx_r}{dt} = r_x x_0(t) - bx_r(t)$$
(5.37)

Because Eqn. (5.35) involves the current time $t$ and the previous time $t - \tau$ at which compromised cyber assets were destroyed, the time dependence of the state variables is explicit on the right sides of these equations. We now have a differential delay equation. Such equations often occur in population biology (MacDonald 1989) and, we now see that they naturally arise in the study of cyber system variability. One property of differential-delay equations is that delays can introduce oscillations in systems that will otherwise not have them. We will briefly explore that idea here and in more detail in the next chapter.

When $a_{co}$ is less than the threshold value for the persistence of compromise, as in Figure 5.19, we expect the extinction of compromise so that all cyber assets are uncompromised in the steady state. As the delay in replacing destroyed cyber assets increases, the number of uncompromised cyber assets (solid black line) falls further below the number of uncompromised cyber assets when none are destroyed (dotted black line). I leave it to you to either imagine or compute the differences in performance, but note that recovery is complete. Although I have not shown them here, as $\tau$ declines the dotted and solid black trajectories become closer and closer. Whether a delay of 30–90 time units is operationally sensible is not a question that we can answer in the abstract, but spurs thinking about the location of the replacement pool relative to the focal cyber system.

When $a_{co}$ is greater than the threshold for persistence of compromise, as in Figure 5.20, in the steady state the same level of uncompromised and compromised cyber assets is reached, regardless of the value of the delay. Note however, (i) that the distance between

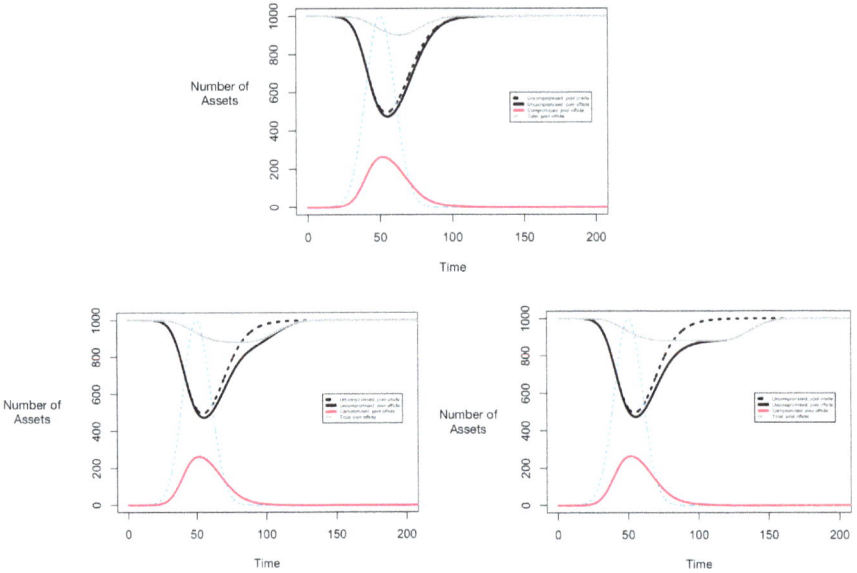

**Fig. 5.19.**    The dynamics of cyber assets during a pulse attack in which $a_{co}$ is less than the threshold for the persistence of compromise after the attack, 10% of assets are destroyed during the attack and have to be replaced from a pool that is 30, 60, or 90 time units away (upper panel, lower left, and lower right panels, respectively). The pulse is shown as a thin dotted line, the dynamics of uncompromised cyber assets in the absence of destruction as a black dotted line, the number of uncompromised, compromised, and total cyber assets as solid black, red, and gray lines, respectively.

the dotted and solid black lines is less than in Figure 5.19 (why is that – think about the role of $a_{co}$), and (ii) after the pulse attack ends, the solid black line oscillates around the dotted black line. One interpretation is that co-compromise is more dominant than destruction of uncompromised cyber assets. This conclusion will depend on the values of $a_{co}$ and $f_d$ and would be an interesting sensitivity study.

We can understand oscillations in terms of the processes of destruction and compromise of cyber assets: cyber assets that are destroyed during the pulse attack are replaced at the same rate at which they are destroyed but unlike the previous situation, when $a_{co}$ is less than the threshold for the persistence of compromise, some of the replacement cyber assets are co-compromised after they enter the cyber system. This leads to the number of uncompromised cyber

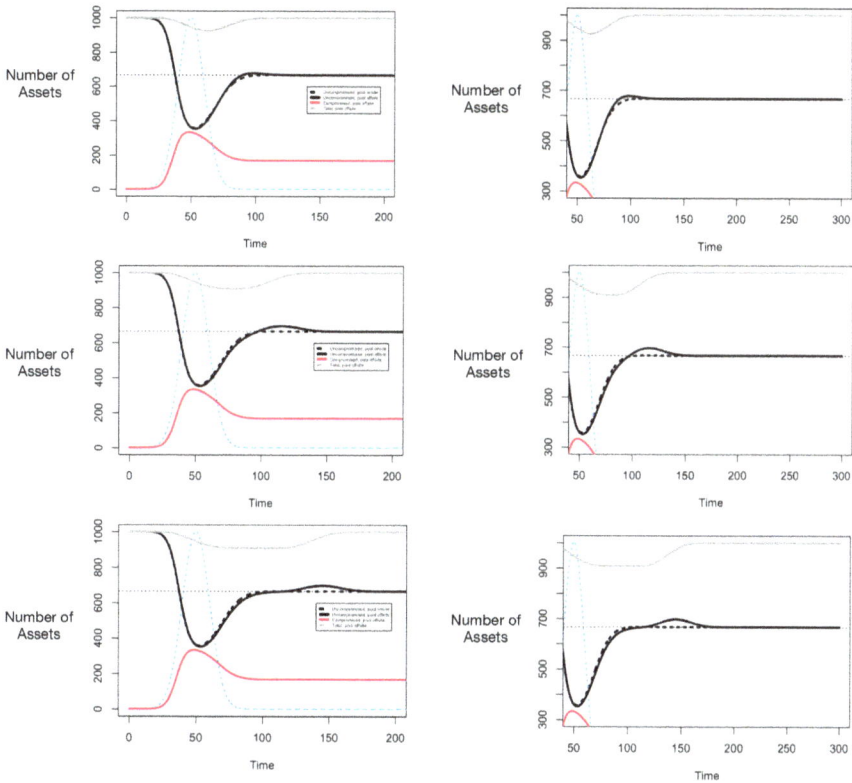

**Fig. 5.20.** The dynamics of cyber assets during a pulse attack in which $a_{co}$ is greater than the threshold for the persistence of compromise, 10% of assets are destroyed during the attack and have to be replaced from a pool that is 30, 60, or 90 time units away (upper, middle and lower panels, respectively). The pulse is shown as a thin dotted line, the dynamics of uncompromised cyber assets in the absence of destruction as a black dashed line, the number of uncompromised, compromised, and total cyber assets as solid black, red, and gray lines, respectively. The left panels can be compared to the three panels in the previous figure. The right panels show expanded time and number scales to give a better sense of the oscillations.

assets when destruction occurs to be temporarily greater than the number in the absence of destruction of cyber assets during the attack. Ultimately, all cyber assets are replaced so that the gray lines in Figure 5.20 approach the initial number of uncompromised cyber assets, but now there is a mixture of compromised and uncompromised assets in the steady state.

> **Potential project**: The oscillations in Figure 5.20 are relatively small compared to the steady state value, so we might just think of them as some kind of unexplained "noise", but to paraphrase Nero Wolfe, in a world of cause and effect all explanations attributed to noise should be approached carefully. Code this model, and explore how other choices of parameter values affect the size of the oscillations.

## 5.7. Allocation of Resources to Defense, Detection of Compromise, and Restoration from Compromise, Under a Constraint on Total Resources

In Chapter 2, we considered a design tradeoff between the rate of detection of compromise $r_x$ and the rate resetting $b$ of compromised cyber assets to the uncompromised state. In this section, we expand on the design tradeoffs on restoring and attack rate parameters from Chapter 2 to think about how to design cyber systems that are Flexible, Adaptive, and Robust (FAR).

Previously, we envisioned resource allocation to be captured in the parameters $b$ and $r_x$ of Eqns. (5.1) and (5.2), and explored the situation in which their sum was constrained to be a constant. We now expand the resource space to include a resource $c$ that can be used to reduce the rate of compromise during attack. In particular, we assume that the rate of attack is now $\frac{axI(t)}{1+\gamma c}$, where $\gamma$ is a parameter chosen so that $c$ has the same units as $b$ and $r_x$. We assume that resources are constrained in the sense that $c_b b + c_r r_x + c_c c \le \mathcal{R}$, where $\mathcal{R}$ is the limit on resources and $c_b, c_r$, and $c_c$ are the unit costs of the relevant parameters $b, r_x, c$ respectively. As before, for simplicity these costs are all set to 1 (I leave alternatives for you to explore).

For example, suppose that we are thinking about how to construct a Cyber Protection Team. We can allocate people and equipment to (i) reduce the rate of attack ($c$), (ii) detect compromised cyber assets and move them to the restoring pool ($r_x$), or (iii) restore compromised assets to operational status ($b$). How does performance depend on our choice of allocation, given the resource constraint?

We could approach this as an optimization problem, asking for the optimal allocations of resources. Such a problem can be solved by the method of dynamic programming (e.g. Dixit and Pindyck 1994, Bertsekas 1995, Mangel 2015) in its deterministic or stochastic versions, depending upon which dynamics we assume for the cyber system, or approximate dynamic programming (Powell 2011, Hackett and Bonsall 2018, 2019).

However, in order to assess the flexibility, adaptive nature, and robustness of allocation decisions, we want more than points that correspond to the optimal allocations. To do so, we need to more fully explore the relationships between performance and allocations. We continue to use the performance functions in Figure 5.13.

When we include resource allocation to reduce the rate of compromise, Eqns. (5.1) and (5.2) become

$$\frac{dx}{dt} = -\frac{axI(t)}{1 + \gamma c} - a_{co}xx_0 + b(X_T - x - x_0) \qquad (5.38)$$

$$\frac{dx_0}{dt} = \frac{axI(t)}{1 + \gamma c} + a_{co}xx_0 - r_x x_0 \qquad (5.39)$$

For the numerical exploration, I let $c$ range from 0 to $0.95\mathcal{R}$ in 25 steps. (As an exercise: can you explain why it is sensible to set $c = 0$ but not sensible to set $c = \mathcal{R}$?) I then considered 9 splits of the remaining resources, $\mathcal{R} - c$, to $b$ and $r_x$, letting the ratio $b : r_x$ range from 10%:90% to 90%:10%, i.e. $b$ was a fraction $0.1, 0.2, \ldots, 0.9$ of the remaining resources and $r_x$ the rest. I set $\gamma = 50$ for computations.

In the following results, I show line graphs of the minimum and quasi-steady state values of $x(t)$ ($y$-axis) as a function of the fraction of resources allocated to reducing the rate of compromise ($x$-axis), $c/\mathcal{R}$, and color code the lines for the split of $\mathcal{R}$ between $b$ and $r_x$. I show the minimum and quasi-steady state values of $x(t)$ using heat maps in which the $y$-axis is the fraction of resources allocated to defense, $c/\mathcal{R}$ and the $x$-axis is the percentage of the remaining resources, $\mathcal{R} - c$, allocated to restoring (i.e. $b$).

### 5.7.1. $a_{co}$ *less than the base case threshold for the persistence of compromise*

In Figure 5.21, I show the minimum and quasi-steady states of the number of uncompromised cyber assets. For both minimum and steady state values, there is a broad range of values of $c$ for which the minimum or quasi-steady state is approximately constant as long as

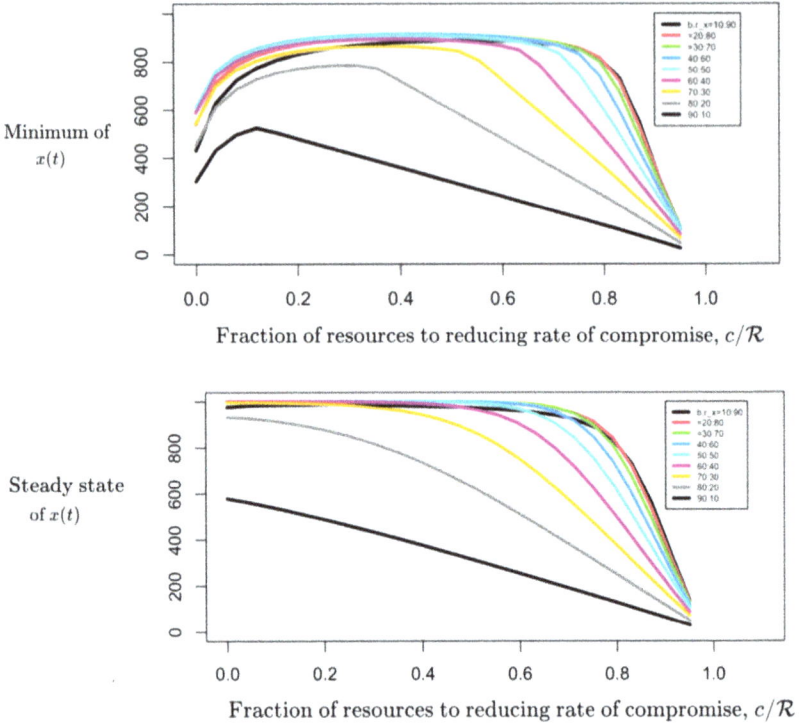

**Fig. 5.21.** Upper panel: The minimum number of uncompromised cyber assets $x(t)$, when $a_{co}$ is less than the base case threshold for the persistence of compromise, as a function of the fraction of resources allocated to reducing the rate of compromise by external attack $c/\mathcal{R}$ for the nine different allocations of the remaining resources $\mathcal{R} - c$ to $b$ and $r_x$. Lower panel: The quasi-steady state number of uncompromised cyber assets $x(t)$, when $a_{co}$ is less than the base case threshold for the persistence of compromise as a function of the fraction of resources allocated to reducing the rate of compromise due to external attack $c/\mathcal{R}$ for the nine different allocations of the remaining resources $\mathcal{R} - c$ to $b$ and $r_x$.

$b$ does not get too big. For fixed $c$, as the allocation of the remaining resources $\mathcal{R} - c$ towards $b$ increases, both minimum and steady state values of the number of uncompromised resources decline, which we can interpret as investing too much in restoring from compromise and not enough in detection of compromise. The declines of all the curves as $c/\mathcal{R}$ approaches the maximum value of 0.95. Is this consistent with your answer to the question I posed above on the range of $c$?

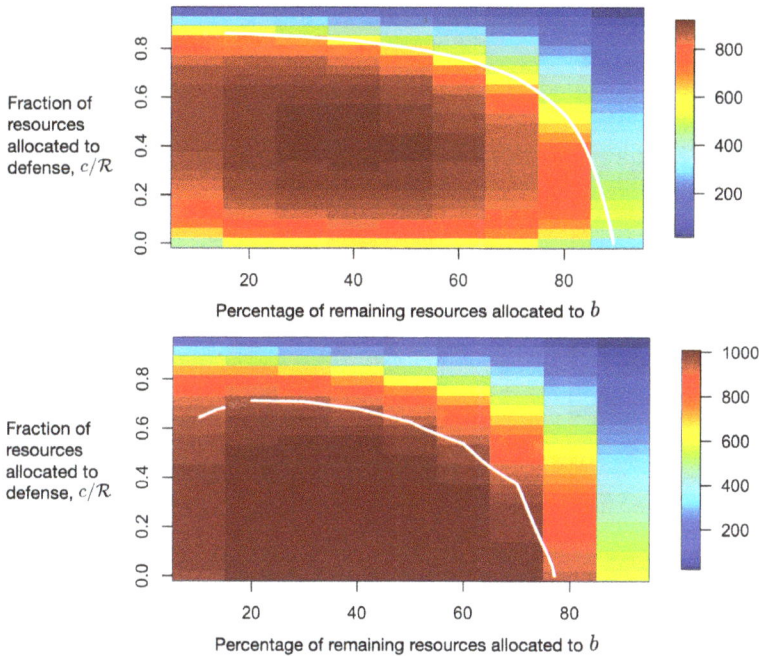

**Fig. 5.22.** Upper panel: The minimum number of uncompromised cyber assets, when $a_{co}$ is less than the base case threshold value for the persistence of compromise as a function of the fraction of resources allocated to reducing the rate of compromise $c/\mathcal{R}$ for the 9 different allocations of the remaining resources $\mathcal{R}-c$ to $b$ and $r_x$. The white line shows the contour of 600 uncompromised cyber assets. Lower panel: The quasi-steady state number of uncompromised cyber assets and the contour of 950 uncompromised cyber assets. Note the difference in the scales of the two maps.

In Figure 5.22, I use a heat map to illustrate the same ideas, as well as contours for minimal and quasi-steady state numbers of cyber assets. Once again we see the very broad and flat regions of minimal and steady state numbers of uncompromised resources. Clearly, there is an optimal combination of parameters that maximizes both the minimum number and the steady state number of uncompromised cyber assets, but the "very good" number region is broad and wide.

In Figures 5.23 and 5.24, I show the minimum and quasi-steady state performance for the four values of the dispersal parameter $\sigma$ in the performance function, and contours of 80% minimum performance, and 90% steady state performance.

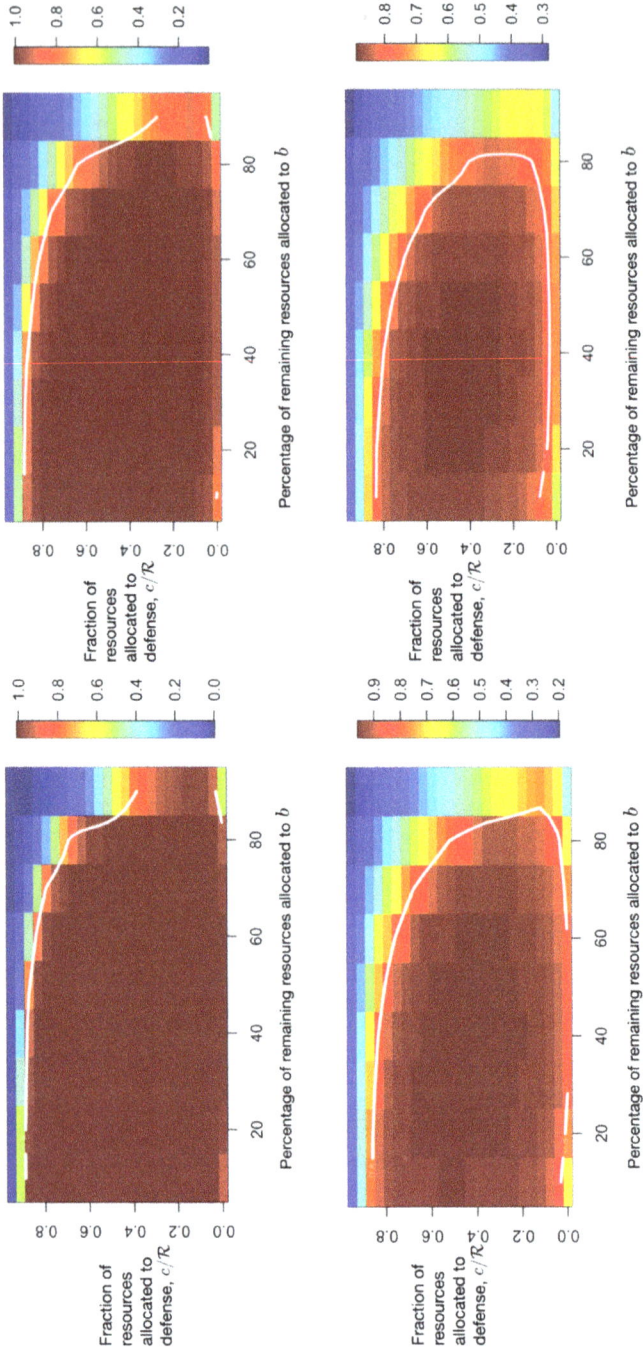

**Fig. 5.23.**    Minimum performance for $\sigma = 50, 100, 200,$ and $300$ (upper left, upper right, lower left, and lower right panels, respectively), when $a_{co}$ is less than the base case threshold for the persistence of compromise, as a function of the fraction of resources allocated to reducing the rate of compromise $c/\mathcal{R}$ for the nine different allocations of the remaining resources $\mathcal{R} - c$ to $b$ and $r_x$, with the contour line showing 80% performance. Note the difference in the scales of the upper and lower heat maps.

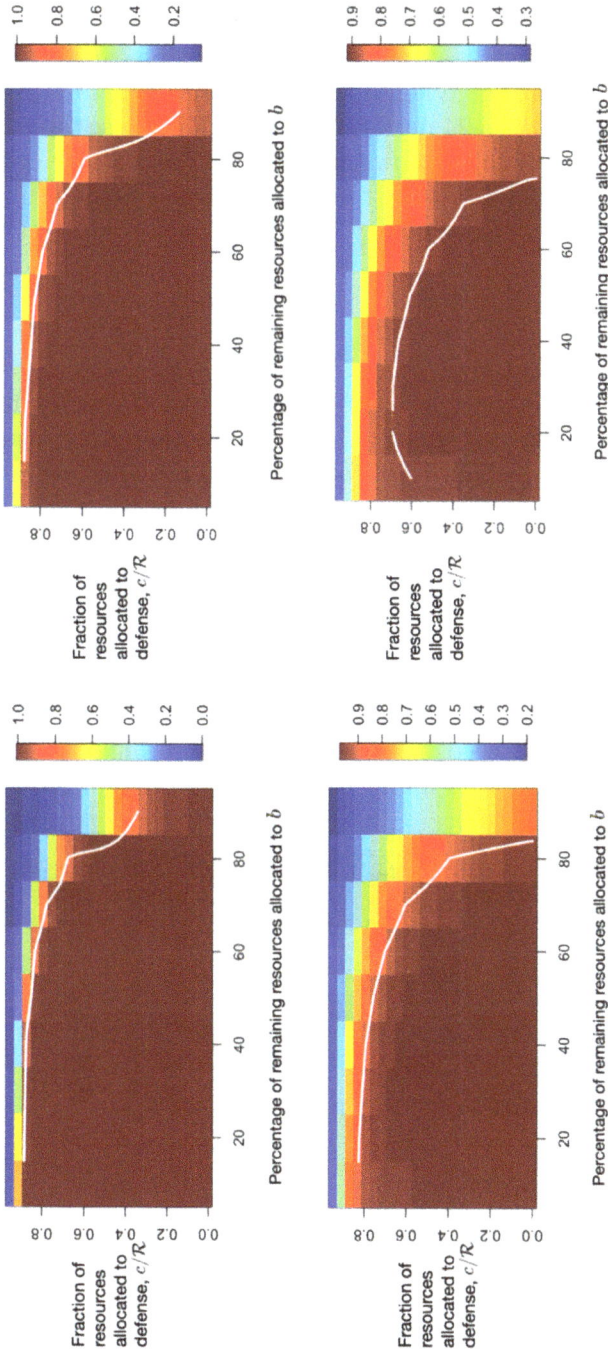

**Fig. 5.24.** Quasi-steady state performance for $\sigma = 50, 100, 200,$ and $300$ (upper left, upper right, lower left, and lower right panels, respectively), when $a_{co}$ is less than the base case threshold for the persistence of compromise, as a function of the fraction of resources allocated to reducing the rate of compromise $c/\mathcal{R}$ for the nine different allocations of the remaining resources $\mathcal{R} - c$ to $b$ and $r_x$, with the contour line showing 90% performance. Note the difference in the scales of the upper and lower heat maps.

We again see very broad regions of good performance – both at the minimum and the steady state – but we also to be careful when thinking about generalizing. When $\sigma = 50$, as in the upper left hand panels of Figures 5.23 and 5.24 performance is high because the number of cyber assets never falls below the midpoint ($x_{50} = 300$) of the performance function. Clearly the story would change with a different midpoint value (a good sensitivity analysis for you to do). At the other limit, when $\sigma_x = 300$, the contour of 90% steady state performance makes clear which combinations of parameters will deliver good performance and which will fail to do so.

When $a_{co}$ is less than the threshold for persistence of compromise, compromise will be extinguished and the system will recover to a fully uncompromised steady state. This suggests that we consider how the recovery time, which I will define to be the time to reach $0.97X_T$ depends on the parameters. But there is a subtlety: The threshold value for compromise is $a_{co_{thr}} = r_x/X_T$ and as we sweep over allocations to $b$ and $r_x$ when $a_{co}$ is fixed it may exceed $r_x/X_T$, leading to persistence of compromise. Thus, we expect that there will be parameter combinations in which recovery does not occur.

As shown in Figure 5.25 there is a large region of the plane where recovery does not occur at all, because the allocations to $c$ and $b$ mean that $a_{co}$, which is fixed, exceeds $r_x/X_T$ and that compromise persists. To emphasize this point, in Figure 5.26, I show the trajectories of

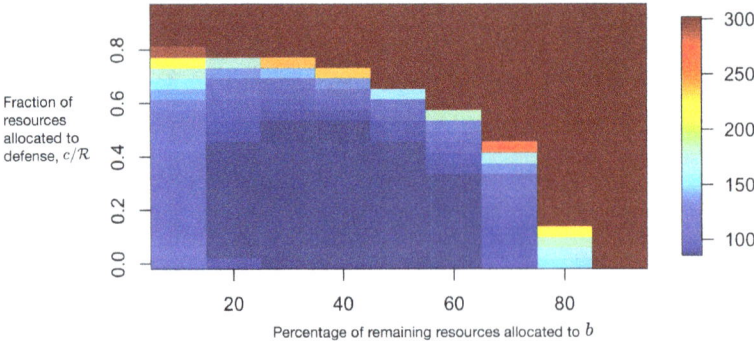

**Fig. 5.25.** Recovery time as a function of the fraction of resources allocated to reducing the rate of compromise $c/\mathcal{R}$ for the nine different allocations of the remaining resources $\mathcal{R} - c$ to $b$ and $r_x$, with the contour line showing 90% performance. The broad brown region at $t = 300$ corresponds to reaching the time horizon without any recovery.

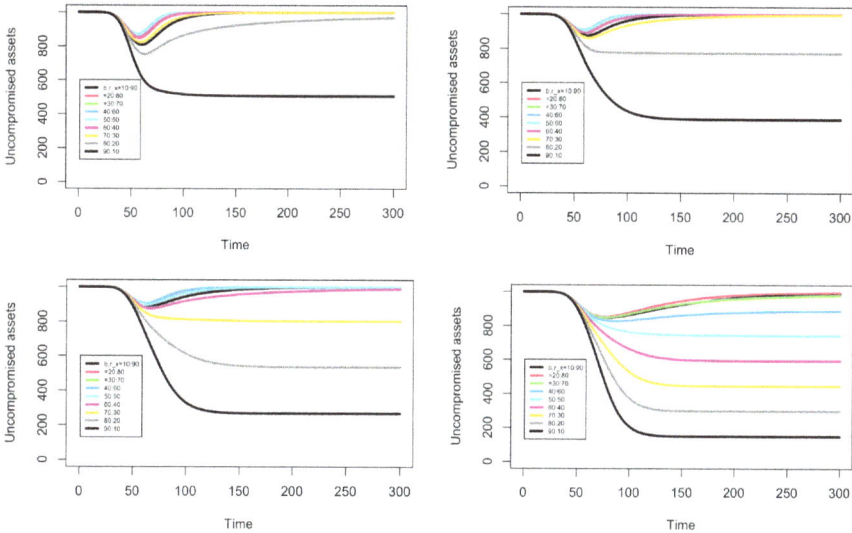

**Fig. 5.26.** Trajectories of the number of uncompromised cyber assets during and after the pulse attack, for allocation to defense $c = 0.095$, $0.21$, $0.33$, or $0.45$ when $\mathcal{R} = 0.6$ for all nine splits between $b$ and $r_x$. Even for the smallest value of $c$, the 90% (to $b$)-10% (to $r_x$) split of resources leads to persistence of compromise after the pulse ends, for reasons discussed in the text. This holds true even when $c = 0$.

$x(t)$ as the allocation to $c$ is about 15.8%, 35.6%, 55.0%, or 80% of $\mathcal{R}$. Even when $c$ is small (or even 0, as it was in Chapter 2), an allocation of resources that is too heavily weighted to restoration ($b$) leads to the persistence of compromise. As the allocation of resources to reducing the rate of compromise increases, the constraints on the $b : r_x$ allocation become more restrictive.

We conclude that the seemingly good idea of using resources to reduce the rate of compromise requires concomitant attention to the allocations of resources to the detection of compromise and restoration of compromised resources.

### 5.7.2.   $a_{co}$ greater than the base case threshold for the persistence of compromise

When $a_{co}$ is greater than the base case threshold value for the persistence of compromise, we expect that the effects described in the

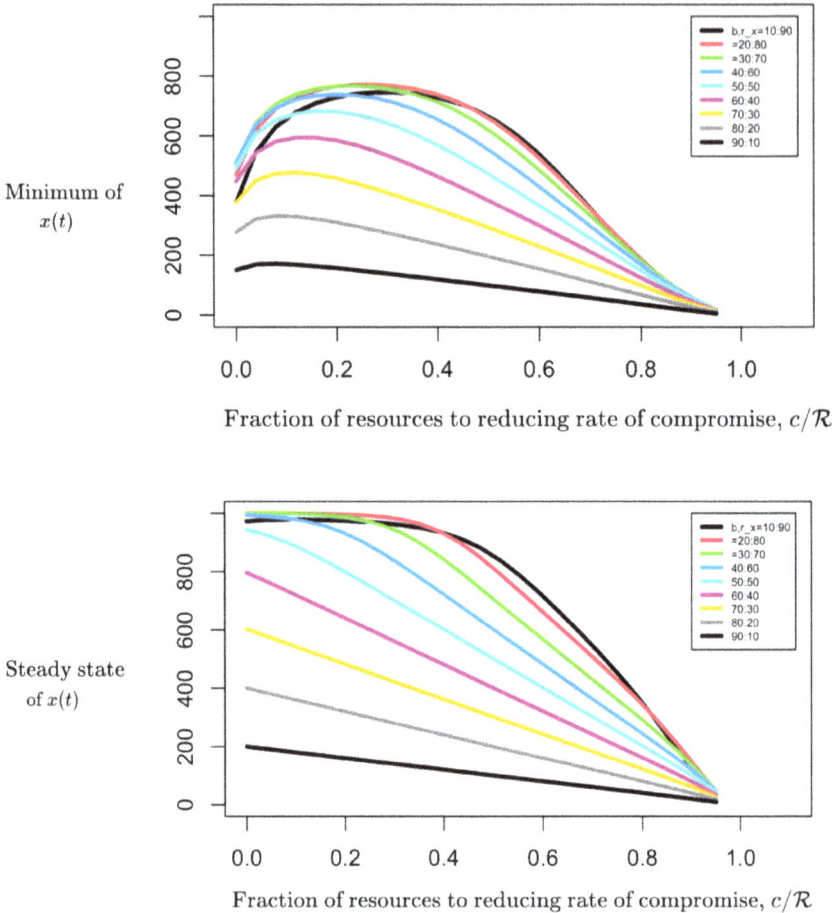

**Fig. 5.27.** The comparison figure for Figure 5.21 when $a_{co}$ exceeds the base case value for the persistence of compromise.

previous section will be amplified. We find

- Line graphs of the minimum and steady state numbers of uncompromised cyber assets (Figure 5.27) have narrower regions of high numbers of uncompromised cyber assets.
- Heat maps of the minimum and quasi-steady state numbers of uncompromised cyber assets (Figure 5.28) have more constricted regions of a high number of uncompromised cyber assets.

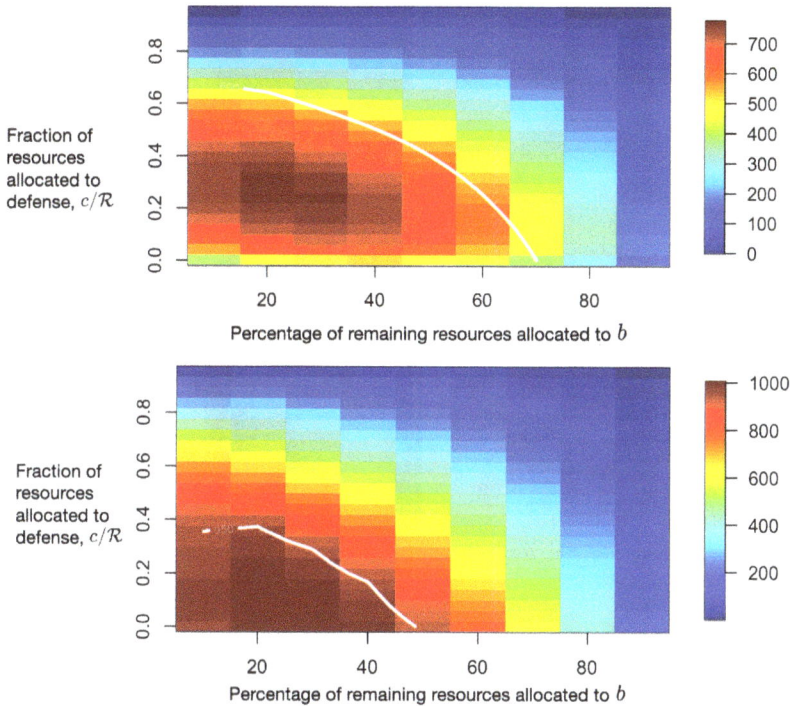

**Fig. 5.28.** The comparison figure for Figure 5.22 when $a_{co}$ exceeds the base case threshold for the persistence of compromise.

- Heat maps of minimum performance (Figure 5.29) have a much more constrained regions of acceptable performance, and the contours shrinks more rapidly than when $a_{co}$ is less than the base case threshold for the persistence of compromise.
- Heat maps of quasi-steady state performance (Figure 5.30) show similar shrinkage of the contour of acceptable steady state performance.

---

**Potential project**: These summary points may make you think of the constraint surface in three dimensions: $b, c$, and $r_x$. If you like making three-dimensional graphs, give it a try!

---

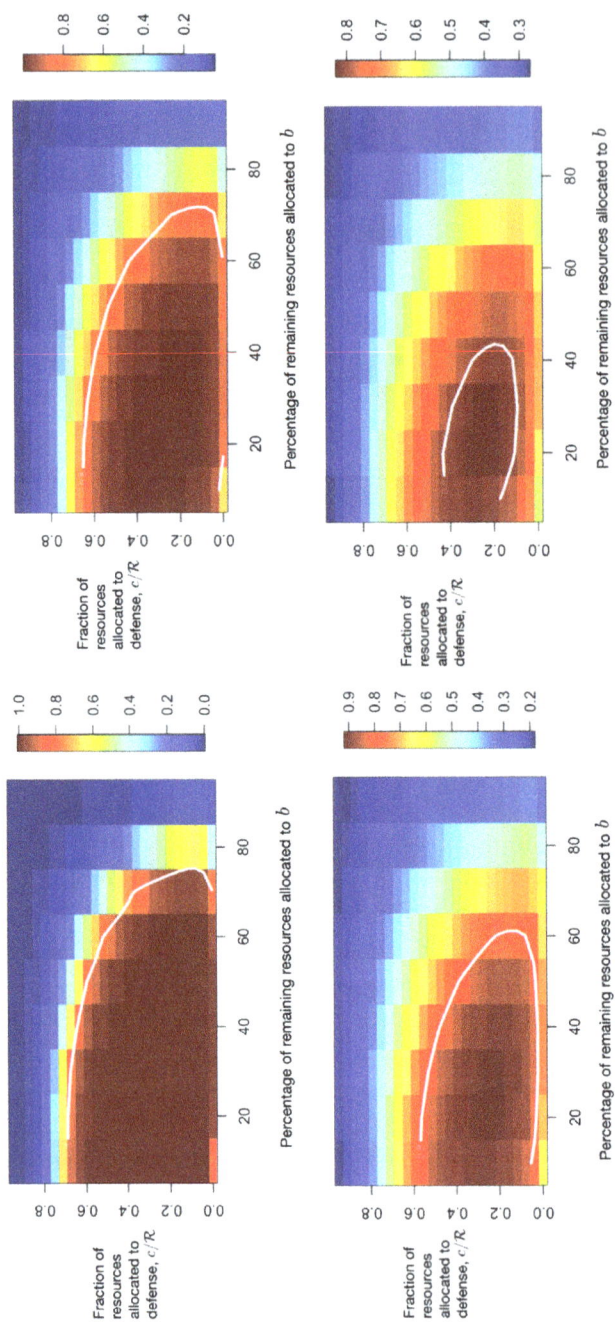

**Fig. 5.29.**  The comparison figure for Figure 5.23 when $a_{co}$ exceeds the base case threshold for the persistence of compromise. Note the difference in the scales.

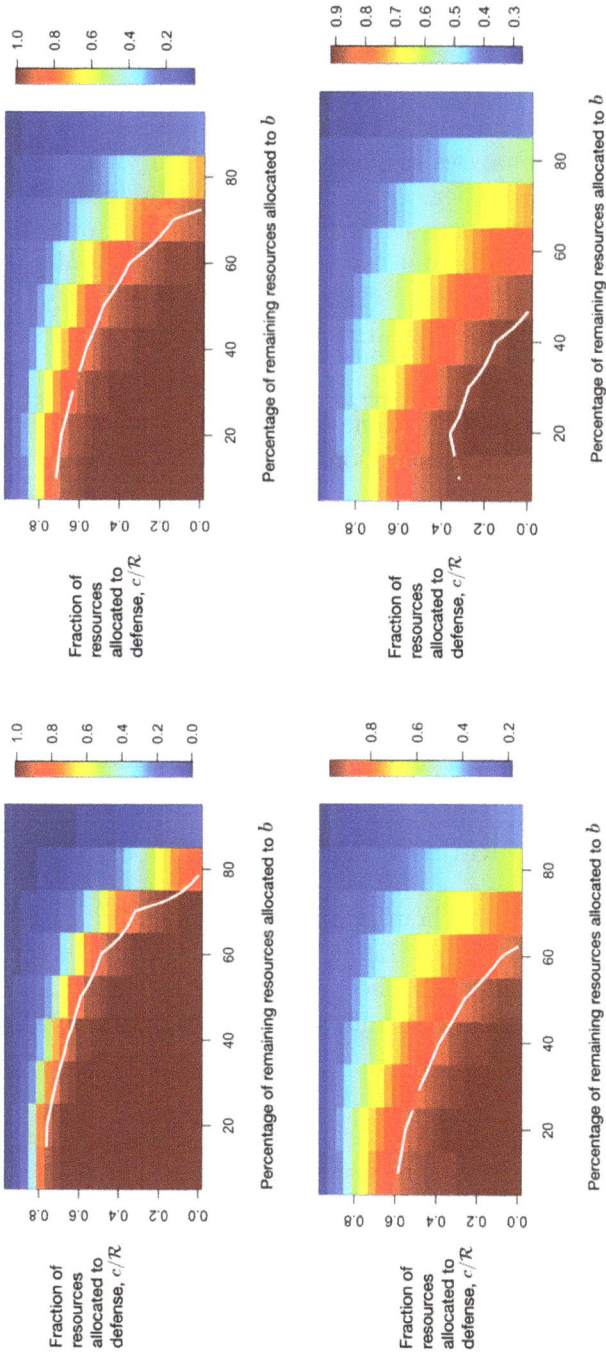

**Fig. 5.30.** The comparison figure for Figure 5.24 when $a_{co}$ exceeds the base case threshold for the persistence of compromise. Note the difference in the scales.

## 5.8.    Adaptation of the Performance Function

In the resilience stack (Figure 1.4), adaptation of the cyber system is the deepest level of resilience. Throughout this book, we conducted various sensitivity analyses of performance by varying the parameters $x_{50}$ and $\sigma_x$ of the performance function. For example, in this chapter for sensitivity analyses we used four values for $\sigma_x$, but we kept $x_{50}$ constant throughout, recognizing that many of the quantitative results would change with a different value of $x_{50}$ (which is why I suggested it as another worthwhile sensitivity analysis).

> **Potential project**: Keeping the sigmoidal form in Eqn. (5.4), envision an adaptation in which $\sigma_x$ and $x_{50}$ are resources and that adaptation involves them changing within a resource constraint. The constraint equation now becomes $c_b b + c_r r_x + c_c c + c_{x_{50}} x_{50} + c_{\sigma_x} \sigma_x \leq R$. Explore the dynamics of cyber assets and performance and characterize the broad tradeoffs in this five-dimensional parameter space. Then imagine the cyber system has experienced a pulse attack and afterwards the performance function is to be adapted. How would you do this?

## 5.9.    Summary of Major Insights

The PAM is intended as heuristic tool that has much in common with many systems but is not intended to model any specific system. The extensions of the PAM show the power of such an approach for developing understanding and quantitative predictions in cyber systems:

- The equations of the PAM generalize when there are multiple pulse attacks over time. In such a case, one needs to specify the times of the peak, dispersal parameter, and rate parameter of each of the attacks. Even when $a_{co}$ is less than the threshold for the persistence of co-compromise, the cyber system may not recover fully and whether it does or not depends upon the timing and intensity of the pulses.

- By including the probability that the defender initiates either a kinetic attack or a cyber attack on critical civilian infrastructure, the attacker can use the PAM to determine the attack rate parameter $a$ that is consistent with a targeted reduction in performance of the defender's cyber system or enabled physical system but below a threshold value for the probability that the defender responds with a kinetic attack or attack on critical civilian infrastructure. The PAM thus becomes a planning tool for the attacker.

- When a CPTs is required for restoration of compromised cyber assets, another time dependency is introduced into the equations for the PAM. Regardless of whether the CPTs visit on a regular schedule or according to threshold number of compromised assets, the PAM generalizes directly, and should stimulate research by the defenders about the operation and effectiveness of CPTs.

- A straightforward extension of the PAM allows us to consider situations in which cyber assets that are restored to uncompromised status can be temporarily hardened to cyber attack, losing that defense over time. None of the qualitative conclusions based on the basic PAM, particularly concerning the role of the threshold level of co-compromise for the persistence of compromise in the steady state, change.

- A straightforward extension of the PAM allows us to consider situations in which assets are divided into those critical for the performance of the cyber system or the enabled physical system and those that have secondary roles, such as reducing the rate of attack on the critical cyber assets. In this case the number of differential equations in the model expands because we must track the dynamics of the two kinds of cyber assets. This extension allows us to study the role of protection of critical assets by secondary cyber assets. In particular, we can explore how performance of the cyber system or enabled physical system is shaped by the parameters characterizing the protection provided by secondary cyber assets and the rate at which critical cyber assets are returned to operational status.

- When cyber assets may be destroyed during an attack, the total number of cyber assets is no longer constant and we must make an assumption about the way destroyed cyber assets are replaced (or not). When destroyed cyber assets are not replaced so that the total number of cyber assets declines, the parameters of the performance function interact with the probability of destruction of an cyber asset during attack to determine how much steady state performance is degraded by the loss of cyber assets. When destroyed cyber assets are replaced from an on-hand pool of uncompromised cyber assets, the extension of the PAM allows us to determine the size of the reserve pool to maintain a sufficient level of performance. When destroyed cyber assets are replaced from an off-site pool of uncompromised cyber assets, a delay (the time for uncompromised replacement cyber assets to reach the cyber system) is introduced into the equations for the PAM. Such a delay can lead to oscillations into the dynamics of uncompromised cyber assets long after the pulse attack has ended.

- To design cyber systems that are Flexible, Adaptive, and Robust we can envision the parameters of the PAM, particularly the rates at which resetting cyber assets are returned to uncompromised states and at which compromised cyber assets are moved from the compromised pool to the resetting pool, as design parameters with a total resource constraint. Furthermore, we can add a third resource whose role is to reduce the rate of external attack. These considerations lead to a straightforward extension of the PAM. Although optimization of steady state dynamics or performance is clearly possible, sweeping over parameter values shows the optima for the number of cyber assets in the steady and performance and that a broad range of values that are close to optimum. That is, the surfaces characterizing the minimum and steady state levels of performance are relatively flat around the peak. The surfaces are broader when rate of co-compromise is less than the threshold for the persistence of co-compromise.

- Ultimately, one may choose to adapt the performance function in response to or anticipation of cyber attack. In this case the parameters of the performance function can be combined with those characterizing the dynamics of the cyber system to allow analysis of the tradeoff between the dynamics of the cyber system and performance of the cyber system or the enabled physical system.

# Chapter 6

# Extensions of the Fundamental Model of Simultaneous Cyber Operations

*[One could]... think of the rain of cyberattacks like rain itself, something that cannot be stopped by any conceivable means but the damage from which can be reduced. Coming inside when it rains, covering what may be damaged if it gets wet, fixing holes, as well as building soundly and away from floodplains all can convert what could be a disaster into something merely annoying... Rain has to be endured. It is foolish not to spend a dollar to prevent ten dollars' worth of damage (over the lifetime of such an investment); it is equally foolish to spend ten dollars to save a dollar's worth of damage. The trick, therefore, is to find the lowest-cost approach to dealing with what cannot be entirely avoided*

– Libicki (2016, p. 81)

Because most of the extensions of the FMSCO that we consider are relatively straightforward, the level of detail in this chapter is less than in the previous chapter. By the time we have reached the end of this chapter, we will have developed and gone beyond the model that is the starting point in Mangel and McEver (2021).

As we go forward, there will be times in which we add additional pools of cyber assets; I will sometimes show the appropriate modification of Figure 3.1 and other times leave it to you to sketch it.

Our starting point is the equations of the basic FMSCO

$$\frac{dx}{dt} = -axy + b(X_T - x - x_0) \tag{6.1}$$

$$\frac{dx_0}{dt} = axy - r_x x_0 \tag{6.2}$$

$$\frac{dy}{dt} = -cxy + d(Y_T - y - y_0) \tag{6.3}$$

$$\frac{dy_0}{dt} = cxy - r_y y_0 \tag{6.4}$$

with each adversary having a sigmoidal performance function, denoted by $\phi_x(x)$ and $\phi_y(y)$, respectively. We will assume that operations commence with all cyber assets of both sides uncompromised.

We consider:

- Including co-compromise in Eqns. (6.1)–(6.4).
- Allowing a fraction restored cyber assets to be hardened to cyber attack, becoming more vulnerable over time. This is in direct analogy to Eqns. (5.15)–(5.17) and increases the number of equations in the model.
- Requiring that each adversary needs to commit some of its own cyber assets to holding the opponent's assets in a compromised state. That is, even if an adversary can block the ultimate goal of an attack, if the attacker commits its cyber assets to maintain compromise of its opponent's cyber assets that are not the primary target but are linked to it, an attack can continue. Buchanan (2018 p. 45) gives the example of an attack on the US Chamber of Commerce (USCC) in which the FBI helped the remove intruders from the network, but months later it was discovered that a smart thermostat and a wireless printer in one of the USCC facilities were still communicating with computers in China. In the language of population biology, the smart thermostat and wireless printer were refuges (Gause 1934/2019) for compromise. This extension also expands the number of equations characterizing simultaneous cyber operations. When an adversary's own cyber assets are used to hold the compromised cyber assets of the opponent, we can consider this to be a "latent cyber risk", in that

the adversary already has access to the opponent's cyber system (Danzig 2014).

- Including delays in moving compromised cyber assets to the restoring pool or from the restoring pool to the uncompromised pool. As in the previous chapter, delays may induce oscillations and the question becomes "how big must those delays be in order to noticeably affect the dynamics of the cyber system?"

There is always more that one can do. For example, we could add decoy cyber assets that appear to be high value but in fact have no effect on performance. Such decoys can assist in detection of compromise (Mangel and McEver 2021) and lead to the adversary misusing its cyber assets.

## 6.1. Including Co-compromise

We include co-compromise in Eqns. (6.1)–(6.4) exactly as in the PAM. Thus, we let $a_{co}$ and $c_{co}$ denote the co-compromise rate parameters used to characterize the rate X-side and Y-side cyber assets cause their own uncompromised cyber assets to become compromised and have

$$\frac{dx}{dt} = -axy - a_{co}xx_0 + b(X_T - x - x_0) \qquad (6.5)$$

$$\frac{dx_0}{dt} = axy + a_{co}xx_0 - r_x x_0 \qquad (6.6)$$

$$\frac{dy}{dt} = -cxy - c_{co}yy_0 + d(Y_T - y - y_0) \qquad (6.7)$$

$$\frac{dy_0}{dt} = cxy + c_{co}yy_0 - r_y y_0 \qquad (6.8)$$

We expect that including co-compromise will lead to lower values of the numbers of uncompromised cyber assets, which is something that you should confirm either numerically or analytically.

Furthermore, the ideas that we developed about resilience following the end of the pulse attack apply here if we imagine that at some point the Y-side ends the cyber attack on the X-side, so the

Eqns. (6.5) and (6.6) become

$$\frac{dx}{dt} = -a_{co}xx_0 + b(X_T - x - x_0)$$

$$\frac{dx_0}{dt} = a_{co}xx_0 - r_x x_0$$

I have not numbered these equations because they correspond exactly to the PAM following an attack.

## 6.2.   Cyber Assets are Restored Hardened to Attack (and Lose Hardening Over Time) or Still Vulnerable to Attack

In this case, we generalize Eqns. (5.15)–(5.17) to the FMSCO

$$\frac{dx_h}{dt} = -a\rho_x x_h(y_h + y_v) - a_{co}x_h x_0 - g_x x_h$$

$$+ f_{xh}b(X_T - x_h - x_v - x_0) \tag{6.9}$$

$$\frac{dx_v}{dt} = -ax_v(y_h + y_v) - a_{co}x_v x_0 + g_x x_h$$

$$+ (1 - f_{xh})b(X_T - x_h - x_v - x_0) \tag{6.10}$$

$$\frac{dx_0}{dt} = aI(\rho_x x_h + x_v)(y_h + y_v) + a_{co}(x_h + x_v)x_0 - r_x x_0 \tag{6.11}$$

$$\phi_x(x_h, x_v) = \left[ \frac{1}{1 + e^{\frac{x_{50} - (x_h + x_v)}{\sigma_x}}} \right] \tag{6.12}$$

$$\frac{dy_h}{dt} = -c\rho_y y_h(x_h + x_v) - c_{co}y_h y_0 - g_y y_h + f_{yh}d(Y_T - y_h - y_v - y_0) \tag{6.13}$$

$$\frac{dy_v}{dt} = -cy_v(x_h + x_v) - c_{co}y_v y_0 + g_y y_h$$

$$+ (1 - f_{yh})d(X_T - y_h - y_v - y_0) \tag{6.14}$$

$$\frac{dy_0}{dt} = cI(\rho y_h + y_v)(x_h + x_v) + c_{co}(y_h + y_v)y_0 - r_y y_0 \tag{6.15}$$

$$\phi_y(y_h, y_v) = \left[ \frac{1}{1 + e^{\frac{y_{50} - (y_h + y_v)}{\sigma_y}}} \right] \tag{6.16}$$

There is nothing conceptually or computationally difficult with this extension even though it doubles the number of equations. Visualization of results is more complicated because of the additional dynamic variables; I leave this to you.

## 6.3.  Cyber Assets are Required to Hold the Adversary in a Compromised State

We now assume that when the X-side is successful at compromising the Y-sides's cyber assets, the X-side must commit $\eta_x$ of its own cyber assets per Y-side cyber asset to continue to hold the the Y-side cyber assets in a compromised state. Similarly, the Y-side commits $\eta_y$ of its assets per compromised X-side asset. We thus add two new dynamic variables $x_c(t)$ and $y_c(t)$, which are, respectively, the numbers of the X-side and Y-side cyber assets committed at time $t$ to holding the opponent's cyber assets in a compromised state. For each adversary, uncompromised cyber assets are then of two types: a pool of uncompromised cyber assets that can be used for attack and performance, and a pool of uncompromised cyber assets that is committed to holding the opponent's assets (Figure 6.1).

Including the committed pool means that successful compromise of Y-side cyber assets by the X-side causes a decline in X-side cyber assets at rate $\eta_x c x y$. In addition, when Y-side cyber assets move from the compromised pool to the recovery pool, there is a concomitant increase in X-side cyber assets at rate $\eta_x r_y y_0$. Similar reasoning applies to Y-side assets.

We need to make an additional decision about what happens to cyber assets that are being used to hold the adversary in compromise when the adversary escapes from compromise.

We will assume that when the adversary escapes compromise, the cyber assets that were holding the compromise return to the uncompromised pool. An interesting alternative, which I leave for you to explore, is that the committed cyber assets are returned to the compromised pool (i.e. the adversary both escapes compromise and compromises the assets that were holding it).

Excluding co-compromise, we append dynamics for the committed assets to Eqns. (6.1)–(6.4) and obtain a new six-dimensional

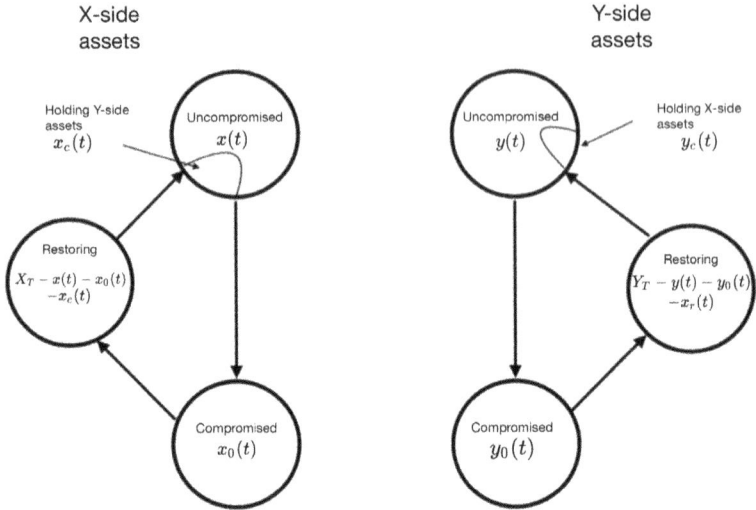

**Fig. 6.1.** When each adversary uses some of its own cyber assets to hold compromise on the other side, the pool of uncompromised cyber assets is composed of a sub-pool that is available for attack and a sub-pool that is uncompromised but not available for attack because it is holding the opponent's cyber assets. To reduce clutter in the figure, I have not shown the rates of compromise, co-compromise, movement from the compromised to recovery pool, or from the recovery pool to the uncompromised pool.

dynamical system

$$\frac{dx}{dt} = -axy - \eta_x cxy + b(X_T - x - x_c - x_0) + \eta_x r_y y_0 \qquad (6.17)$$

$$\frac{dx_c}{dt} = \eta_x cxy - \eta_x r_y y_0 = \eta_x (cxy - r_y y_0) \qquad (6.18)$$

$$\frac{dx_0}{dt} = axy - r_x x_0 \qquad (6.19)$$

$$\frac{dy}{dt} = -cxy - \eta_y axy + d(Y_T - y - y_c - y_0) + \eta_y r_x x_0 \qquad (6.20)$$

$$\frac{dy_c}{dt} = \eta_y (axy - r_x x_0) \qquad (6.21)$$

$$\frac{dy_0}{dt} = cxy - r_y y_0 \qquad (6.22)$$

Note that if we set $\eta_x = \eta_y = 0$ and understand that $x_c(0) = y_c(0) = 0$, Eqns. (6.17)–(6.22) reduce to Eqns. (6.1)–(6.4).

For computations, I used the base case parameters for FMSCO, set $\eta_x = \eta_y = 1$ and assumed that for the X-side performance function $x_{50} = 14$ and $\sigma_x = 4$ and for the Y-side performance function $y_{50} = 65$ and $\sigma_y = 15$.

In Figure 6.2, I show the dynamics of uncompromised and compromised cyber assets in the basic FMSCO on the left panels. I also show their totals $x_{tot}(t) = x(t) + x_0(t)$ and $y_{tot}(t) = y(t) + y_0(t)$. As in Chapter 3, the totals are less than $X_T$ or $Y_T$ because cyber operations are continuous. In the right panels, I show the dynamics

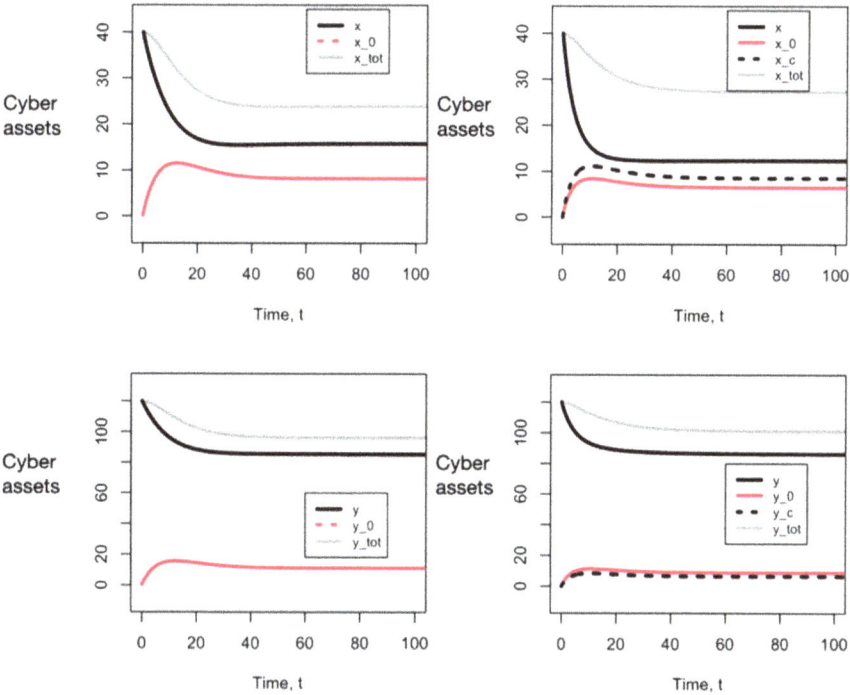

**Fig. 6.2.** Illustrating the effect of having to commit cyber assets to hold compromise in the adversary when $\eta_x = \eta_y = 1$. The upper panels correspond to the X-side cyber assets and the lower panels to the Y-side cyber assets. On the left I show the dynamics of uncompromised cyber assets, compromised cyber assets and their totals $(x, x_0, x_{tot})$ and $(y, y_0, y_{tot})$ respectively in lower panels) for the basic FMSCO. On the right, I show the dynamics when cyber assets ($x_c$ and $y_c$, respectively) are required for holding compromise of the adversary's cyber assets.

of uncompromised, compromised, and committed cyber assets and their totals when cyber assets are required to hold compromise in the adversary's assets.

The totals $x_{tot}(t) = x(t) + x_0(t) + x_c(t) < X_T$ and $y_{tot}(t) = y(t) + y_0(t) + y_c(t) < Y_T$ but note that these totals may be higher than the total for the FMSCO without this extension. This happens because uncompromised cyber assets are freed up when the adversary's cyber assets escape compromise, so holding compromise is in some sense a kind of refuge. Note that because we are including compromised cyber assets in the total, this conclusion would hold even if we assumed that when the adversary escapes compromise it also compromises the cyber assets that were holding it.

The story with performance (Figure 6.3) is also interesting. In the basic FMSCO, the steady state value of the number of

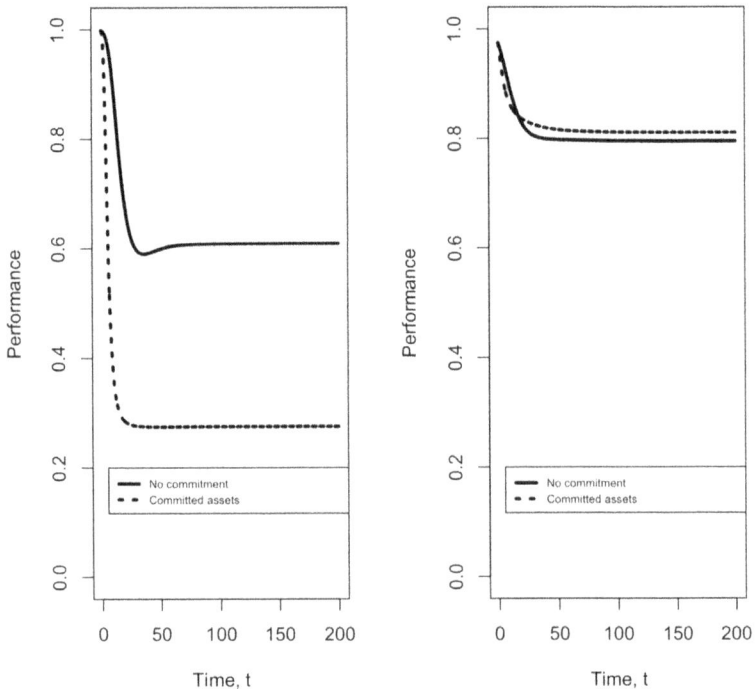

**Fig. 6.3.** The dynamics of performance for the X-side and the Y-side cyber systems or enabled physical systems (left and right, respectively) in the basic FMSCO (solid lines) and when cyber assets are required to hold compromise in the adversary (dotted lines).

compromised assets X-side cyber assets is $\bar{x} = 15.8$, so that performance is considerably reduced (solid line in the left panel of Figure 6.3) but still above 50%, since $x_{50} = 14$. However, because the steady state number of X-side cyber assets that are uncompromised and uncommitted when assets are required to hold compromise in the adversary is $\bar{x} = 10.1$, performance of the X-side cyber or enabled physical system drops considerably (dotted line in Figure 6.3).

The steady states of the Y-side cyber assets are virtually the same whether cyber assets are required for compromise or not. Indeed, the the number of uncompromised cyber assets is slightly higher ($\bar{y} = 86.8$) when cyber assets are required to hold compromise than when they are not ($\bar{y} = 85.3$), because as the X-side cyber assets escape compromise, the Y-side cyber assets holding it are returned to the uncompromised pool. Since $y_{50} = 65$, these steady state values convert to virtually the same performance, with performance slightly higher when cyber assets have to be committed to holding compromise.

As before, we have to be careful not generalize these results but I did not pick the parameters to make them come out this way. However, the message is that (i) we can model the situation when cyber assets are required to hold compromise of the adversary's cyber assets and (ii) it may make a difference for performance of the cyber system or enabled physical system. I leave the sensitivity analysis and parameter sweeps to you.

## 6.4. Delays in Detection of Compromise or Restoration from Compromise

To explore the effects of delays, we return to the basic equations for the FMSCO, and include delays in just the X-side dynamics for simplicity of computation (fewer parameters) and analysis (fewer new equations). We denote the delay in moving cyber assets from the compromised pool to the restoring pool by $\tau_d$ (where the subscript $d$ denotes detection of compromise) and the delay in restoring compromised cyber assets the to uncompromised pool by $\tau_r$.

In this case, we explicitly introduce the number of X-side cyber assets in the recovery pool, $x_r(t)$ but because we continue with the other assumptions of the FMSCO (particularly no destruction of cyber assets during attack) we do not need to introduce the number of Y-side cyber assets in the recovery pool (it is still $Y_T - y(t) - y_0(t)$) so that Eqns. (6.1)–(6.4) become

$$\frac{dx}{dt} = -ax(t)y(t) + bx_r(t - \tau_r) \tag{6.23}$$

$$\frac{dx_0}{dt} = ax(t)y(t) - r_x x_0(t - \tau_d) \tag{6.24}$$

$$\frac{dx_r}{dt} = r_x x_0(t - \tau_d) - bx_r(t - \tau_r) \tag{6.25}$$

$$\frac{dy}{dt} = -cxy + d(Y_T - y - y_0) \tag{6.26}$$

$$\frac{dy_0}{dt} = cxy - r_y y_0 \tag{6.27}$$

We consider both dynamics and the phase plane of the cyber assets. Recall that for the basic FMSCO, the numbers of uncompromised cyber assets declined smoothly in time from $X_T$ and $Y_T$ to their steady state values $\bar{x}$ and $\bar{y}$ and the numbers of compromised cyber assets had transient peaks followed by a monotonic decay to their steady state values $\bar{x}_0$ and $\bar{y}_0$. Furthermore, the phase plane for uncompromised cyber assets had spiral steady state. The phase plane for compromised cyber assets in the FMSCO was essentially a straight line from $x_0(0) = 0, y_0(0) = 0$ that overshot the steady state (because these phase planes are two-dimensional projections of the four-dimensional system that is FMSCO) and then returned to the steady state $\bar{x}_0, \bar{y}_0$ on the same straight line.

### 6.4.1.    *Numerical results*

For computations, I held $\tau_d = 4$ (i.e. 4 days to detect compromise and move a compromised cyber asset to the restoring pool), and let $\tau_r = 4, 8$ or 12 (i.e. 4, 8, or 12 days to restore a cyber asset to the uncompromised state). I used the package *dede* in deSolve to obtain solutions of Eqns. (6.23)–(6.27). In all cases, I assumed that the initial conditions corresponded to only uncompromised cyber assets for each adversary. In Figure 6.4, I show the dynamics and phase planes for uncompromised cyber assets. In Figure 6.5,

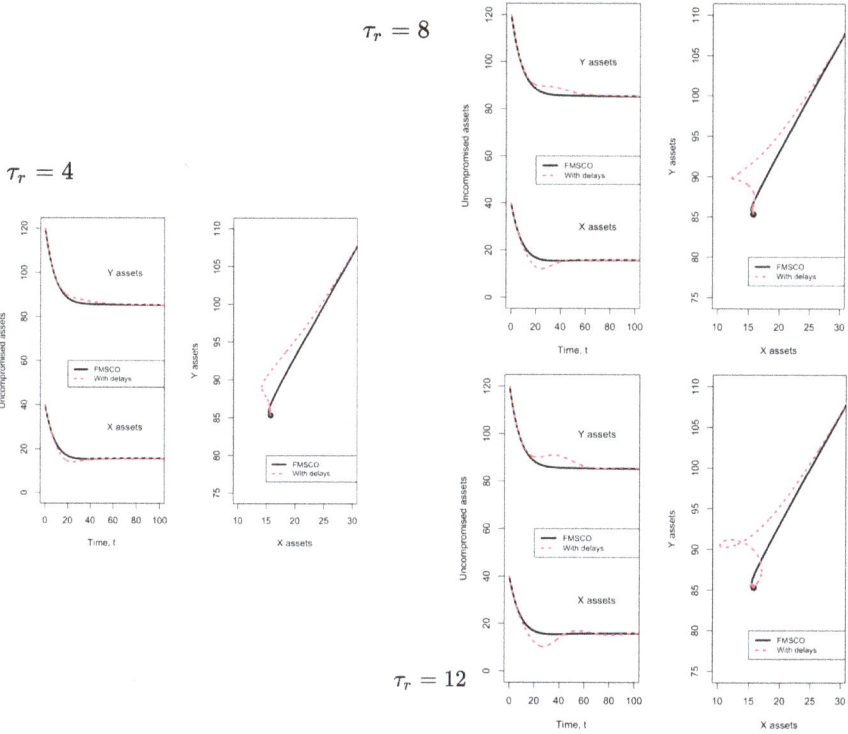

**Fig. 6.4.** The dynamics (left side of each panel) and phase planes (right side of each panel) for uncompromised cyber assets when $\tau_d = 4$ for three values of $\tau_r$. For comparison, the dynamics or phase plane of the basic FMSCO are shown by a solid black line and the dynamics or phase plane of the extension with a delay are shown by the dotted red line.

I show the dynamics and phase planes for compromised cyber assets, respectively.

In Figure 6.4 when $\tau_r = 4$, t the delay in moving to the reset-ting pool is relatively small, so that the dynamics of uncompromised Y-side cyber assets are monotonic and the dynamics uncompromised X-side cyber assets are transitory, which shows in the phase plane as a deviation from the basic FMSCO. When $\tau_r = 8$, the devia-tion in the phase plane from the basic FMSCO is larger but the has the same qualitative features as when $\tau_r = 4$. However, when $\tau_r = 12$, uncompromised cyber assets have notable oscillations and the phase plane trajectory appears to cross itself before spiraling in

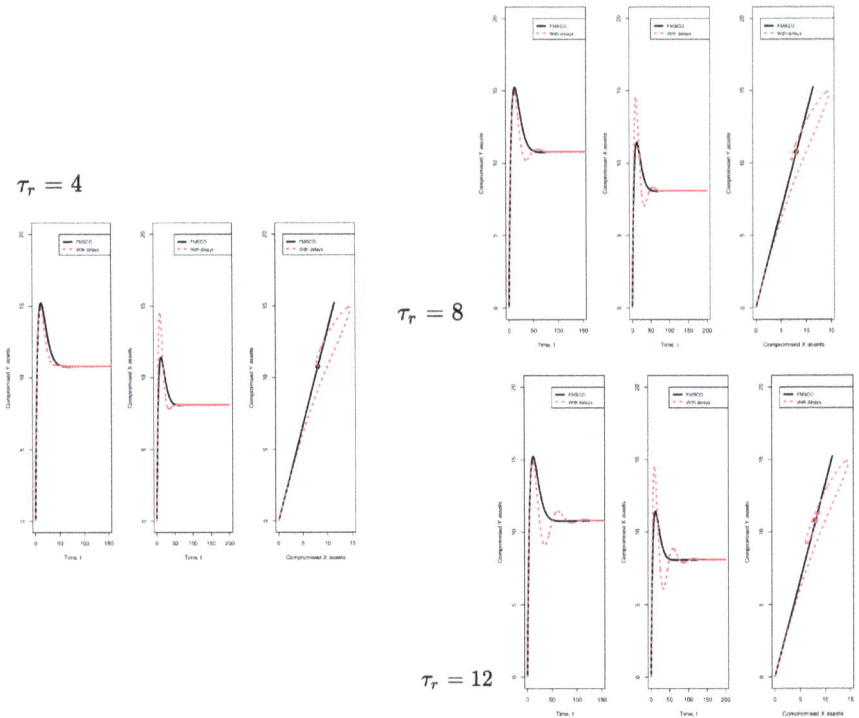

$\tau_r = 4$

$\tau_r = 8$

$\tau_r = 12$

**Fig. 6.5.** The dynamics (two left plots in each panel) and phase planes (right plot in each panel) for compromised assets when $\tau_d = 4$ for three values of $\tau_r$. For comparison, the dynamics or phase plane of the basic FMSCO are shown by a solid black line and the dynamics or phase plane of the extension with a delay are shown by the dotted red line.

to the steady state (this apparent crossing is due to projecting from the four-dimensional phase space to two-dimensional phase plane).

In Figure 6.5, we see that even with the smallest delay, compromised Y-side cyber assets show a transient and compromised X-side cyber assets show a brief oscillation. These dynamics are captured in the phase plane by a large excursion away from the steady state before reaching it. For $\tau_r = 8$, the dynamics of the compromised cyber assets now show (strongly damped) oscillations. In the phase plane, these oscillations are captured by the trajectory briefly cycling around the steady state. For the largest value of $\tau_r = 12$, compromised cyber assets now show stronger oscillations, and the trajectory in the phase plane is clearly that associated with a spiral point.

Why is a calculation like this important? It is almost guaranteed that in operational cyber systems there will be delays in the detection of compromise and restoration to the uncompromised state. We have seen that if the delays are sufficiently large, the number of uncompromised cyber assets can spiral in to the steady state. During that spiral, the number of uncompromised cyber assets may fall below a threshold for a kinetic response or cyber attack on the adversary's critical civilian infrastructure. However, falling below this threshold is only temporary, and a defender may make a different decision about the response given such knowledge.

> **Potential project\***: Conduct a perturbation analysis similar to the one in Chapter 3 to analyze the properties of the steady state when the equations for the FMSCO include delays. There is a rich mathematical literature on differential equations with delays (some starting points are Driver 1977, Cooke and Grossman 1982, Dye1984, Murdoch *et al.* 1987, 2003, Bhunia *et al.* 2023) that may help guide your work.

## 6.5. Summary of Major Insights

- As with the PAM, we have seen how the FMSCO generalizes, which is what makes it a powerful starting point. That is, because the FMSCO is not specific to any particular cyber system but has much in common with many cyber systems and with small modifications can capture other specific situations.
- When X-side cyber assets are used to hold compromise of the Y-side assets, performance of the X-side cyber system or enabled physical system may substantially decline. This is determined by the interaction of the midpoint of the X-side performance function and the number of X-side cyber assets needed to hold Y-side cyber assets in compromise. A similar conclusion applies for Y-side cyber assets holding X-side assets.
- Delays in the detection of compromise (moving compromised cyber assets from the compromised pool to the restoring pool) and restoration (returning cyber assets in the restoring pool to the uncompromised pool) may convert steady states that are

approached monotonically to steady states that are spiral points. In such a case, one sides's uncompromised cyber assets may temporarily fall below a threshold for a kinetic response or an attack on critical civilian cyber systems. This important possibility needs to be clearly communicated to decision-makers.

# Chapter 7

# Including a Distribution in Vulnerability in the Pulse Attack Model

Variation is prominent in biological systems and is the foundation for the theory of evolution by natural selection; indeed variation has been called the core of biology (Berry 1989). In the population biology of disease there is a large literature concerning variation in infectivity of pathogens and in defenses by hosts. A good entry point is Brouwer *et al.* (2019).

Because there is no genetic variation during the production of cyber components, we might think that cyber assets as metaphorical species either have no variation in vulnerability to compromise, or variation that is so small it can be ignored. In some cases, this is probably a very good assumption. In other cases, it might not. For example, Alan Brown (Johns Hopkins University Applied Physics Laboratory) collected data on a network of 7000 computers using the Microsoft Windows Operating System (OS) over the period from July 2019 to March 2020, and identified a subset of 334 computers that appeared consistently in the data, had virtually the same initial OS release (OS build in the Microsoft terminology) but different updates, and did not revert to a previous OS during the monitoring period (details in S1 Appendix in Mangel and Brown 2022).

In Figure 7.1 (upper left panel), I show the results of those analyses. Here we see that the vast majority of the computers were indeed using the most recent OS. Mangel and Brown (2022) then used a

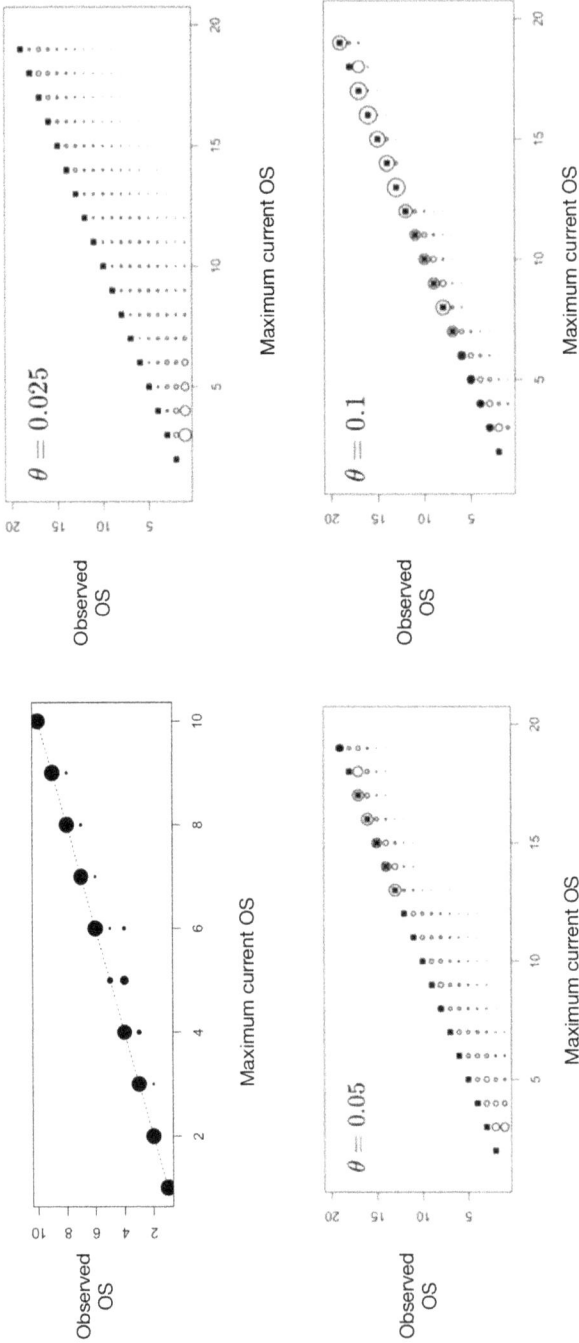

**Fig. 7.1.** Distribution of Operating Systems (OSs) in an actual (upper left panel) and simulated (upper right and both lower panels) cyber systems. In all cases, the the $x$-axis is the most recent OS release and the $y$-axis is the number of computers/cyber assets using a particular OS. The upper left panel was based on about 335 computers from a network of about 7000 computers between July 2019 and March 2020 (Mangel and Brown 2022). I show the relative distribution of the current OS, where the diameter of the circle is proportional to the base-10 logarithm of the counts of that OS in the data. The other three panels show a simulated cyber system in which the probability that a cyber asset is updated is $1 - e^{-\theta}$, so that larger values of $\theta$ indicate higher rates of updates. In each of those panels, squares show the most recent OS, and circles show the fraction of the 1000 simulated cyber assets in the respective OS. For ease of viewing, the fraction is multiplied by 3.

population model in which the probability of updating an OS in a unit interval of time was $1 - e^{-\theta}$, where $\theta$ is a parameter. Three results from Mangel and Brown (2022) are also shown in Figure 7.1. As $\theta$ increases from 0.025 to 0.1 (upper right, lower left, and lower right panels), the representation of less current OS declines, so that the lower right panel looks much like the upper left panel. But we can also easily imagine that the OS of the focal cyber system is not updated rapidly – leading to a wide distribution of OS being used, as in the upper right panel. If the vulnerability of the cyber system to compromise declines as the OS updated, we can easily imagine a distribution of vulnerability.

> **Potential project**: Collect your own data on operating systems in a population of computer users. It does not need to be as complicated as the procedure described in Mangel and Brown (2022). For example, I am a user of the Apple OS, and know lots of people who also are, so I might just ask them what MacOS they use (mine, at this time, is 15.3.2). What inferences can you make about the distribution of variation in operating systems?

Because of such variation, in this chapter we explore what would happen if we included a distribution of vulnerability in the rate of compromise. Since interpretations are easier with the PAM, I focus on it and leave extensions of these ideas to the FMSO to you. Furthermore, we will close with an open problem, which I hope one of you will solve.

## 7.1.  Characterizing the Distribution of Vulnerability

We denote vulnerability by the symbol $v$ with the understanding that cyber assets $x_v$ with vulnerability $v$ are compromised at rate $avx_vI(t)$ during the pulse attack. We assume that $v > 0$, because otherwise the cyber asset would not be vulnerable to attack. The **gamma density** (Mangel 2006) is a very good tool for modeling the distribution of vulnerability among the cyber assets.

There are a number of different formulations of the gamma density; here I show form that is used in R. In this case, we think of vulnerability as a random variable $\tilde{V}$ having gamma density with shape parameter $a_v$ and scale parameter $s_v$ (explained in more detail below) so that

$$\Pr[v \leq \tilde{V} \leq v + dv] = \frac{1}{s_v^a \Gamma(a_v)} v^{a_v-1} e^{-v/s_v} dv + o(dv) \qquad (7.1)$$

where $o(dv)$ represents terms that are higher power in $dv$ and $\Gamma(a)$ is the classical **gamma function** of applied mathematics. One way to think about the gamma function is this. If we let $f(v) = \frac{1}{s_v^a \Gamma(a_v)} v^{a_v-1} e^{-v/s_v}$, since vulnerability has to take some value, $\int_0^\infty f(v)dv = 1$. Consequently, we conclude that $\int_0^\infty v^{a_v-1} e^{-v/s_v} dv = s_v^a \Gamma(a_v)$, so that $\Gamma(a_v) = \int_0^\infty v^{a_v-1} e^{-v/s_v} dv / s_v^a$. Richard Feynman used this property of probability densities to integrate complicated functions in his head (Feynman 1985).

When vulnerability $\tilde{V}$ is characterized by Eqn. (7.1), the mean and variance of vulnerability are $\mathcal{E}(\tilde{V}) = a_v s_v$ and $Var(\tilde{V}) = a_v s_v^2$ so that the coefficient of variation (standard deviation divided by the mean) squared is $CV^2 = 1/a_v$. Try showing this using the iteration property of the gamma function that $\Gamma(z + 1) = z\Gamma(z)$ (also see Mangel 2006 for hints).

We use a discrete distribution for values of vulnerability $v_1, v_2, \ldots, v_N$ and assume that a discrete analogue of Eqn. (7.1) characterizes their probabilities, so that we write

$$\Pr[\tilde{V} = v_n] = v_n^{a_v-1} e^{-v_n/s_v} / \sum_{n'=1}^{N} v_{n'}^{a_v-1} e^{-v_{n'}/s_v} \qquad (7.2)$$

In this equation, I have replaced the normalization constant involving the gamma function in Eqn. (7.1) by its discrete equivalent, using $n'$ as an index so that we do not confuse the numerator and denominator. It is clear that $\sum_{n=1}^{N} \Pr[\tilde{V} = v_n] = 1$ (Figure 7.2).

To simplify writing in the next section, we set

$$f_n = v_n^{a-1} e^{-v_n/s} / \sum_{n'=1}^{N} v_{n'}^{a-1} e^{-v_{n'}/s} \qquad (7.3)$$

so that $\Pr[\tilde{V} = v_n] = f_n$. We refer to the $f_n$ as the initial distribution of vulnerability.

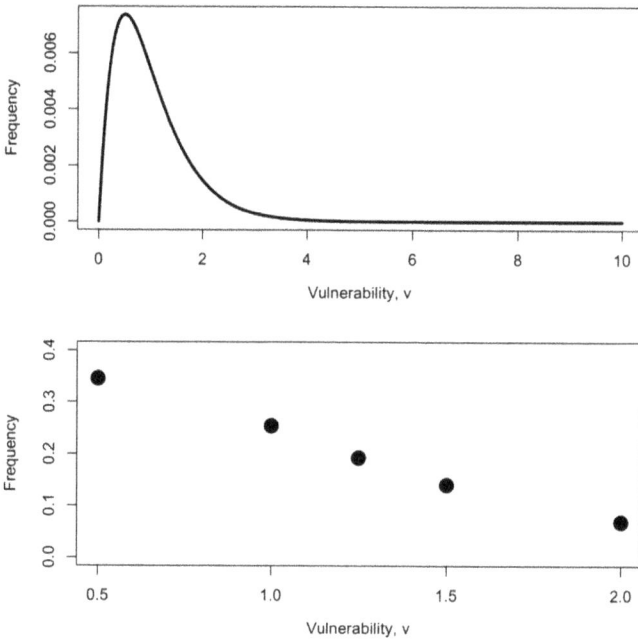

**Fig. 7.2.** Upper panel: The continuous gamma density, corresponding to Eqn. (7.1) with $a_v = 2$ and $s_v = 1/2$ so that the mean vulnerability $\mathcal{E}(\tilde{V}) = 1$. Lower panel: A discrete gamma density for five values of vulnerabilities taking the discrete values $v_n = (0.50, 1.00, 1.25, 1.50,$ and $2.00)$. The mean of this discrete distribution is 1.01.

## 7.2. The PAM with a Discrete Distribution of Vulnerability

We now modify Eqns. (2.2) and (2.3) to take into account a distribution of vulnerability to attack of the cyber assets, in which cyber assets with vulnerability $v_n$ will be compromised at rate $a v_n I(t)$ during the pulse attack. However, we must make additional decisions about what happens after that. Here are the two key questions and the answers that we will use.

(1) How do vulnerability and co-compromise interact? We will assume that vulnerability applies only to external attack, so that there is only one pool of compromised cyber assets and that the mechanism of co-compromise is such that all cyber assets are

equally vulnerable to co-compromise, regardless of how hardened they are to external attack.

(2) After resetting, what is the vulnerability of cyber assets that return to the uncompromised pool? We explore two answers to this question. First, we assume that cyber assets return to the uncompromised pool in proportion to the distribution of initial vulnerabilities $f_n$. Second, we assume that all cyber assets return to the uncompromised pool hardened with minimum vulnerability and then become more vulnerable over time.

If you do not like these choices, that is fine and it means you should change the models given below and explore the your answers.

These assumptions allow us to retain single pools for compromised and resetting assets, and to expand to $N$ pools of uncompromised assets (Figure 7.3).

## 7.3. Formulation of the Dynamics

In light of these assumptions, the generalization of Eqns. (2.2) and (2.3) will have one equation for each of the uncompromised assets, and one equation for the compromised assets.

### 7.3.1. *Cyber assets return according to initial vulnerability*

In this case, the rate of return of cyber assets with vulnerability $v_n$ is $f_n$ times the total rate of return to the uncompromised pool, $b(X_T - \sum_{n=1}^{N} x_n - x_0)$. Thus, the generalization of Eqns. (2.2) and (2.3) is

$$\frac{dx_n}{dt} = -ax_n v_n I(t) - a_{co}x_n x_0 + f_n b\left(X_T - \sum_{n=1}^{N} x_n - x_0\right),$$

$$\text{for } n = 1 \text{ to } N \tag{7.4}$$

$$\frac{dx_0}{dt} = aI(t)\sum_{n=1}^{N} x_n v_n + a_{co}x_0 \sum_{n=1}^{N} x_n - r_x x_0 \tag{7.5}$$

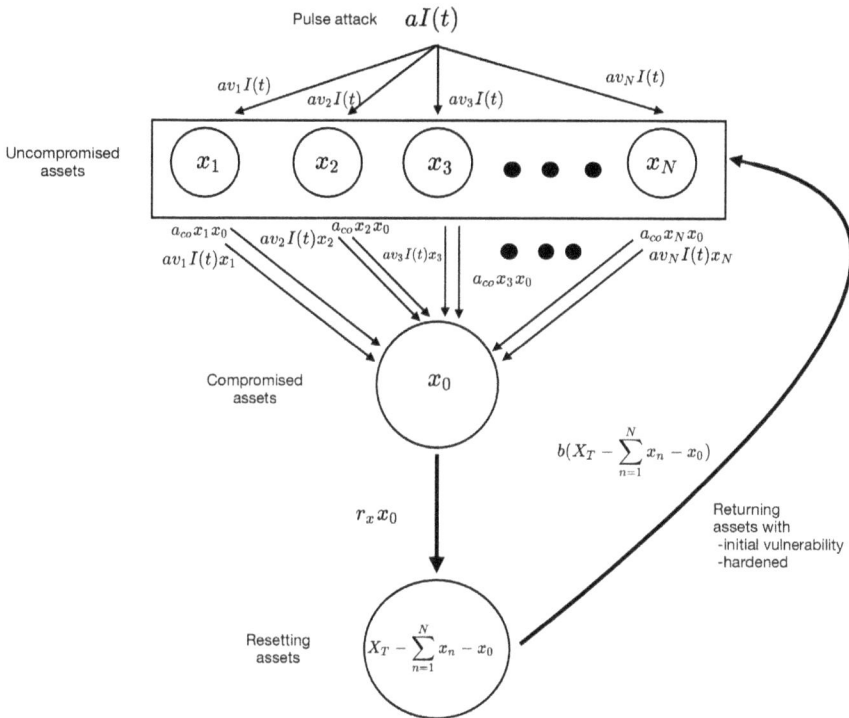

**Fig. 7.3.** When cyber assets differ in the vulnerability to attack, the pool of uncompromised assets is subdivided into the number of assets $x_n$ with vulnerability $v_n$. For simplicity we assume that once compromised, all cyber assets are moved to the compromised pool $x_0$, after which they are able to co-compromise. Compromised cyber assets are moved to the resetting pool and from there back to the uncompromised pool, returning according to the initial distribution of vulnerability or hardened to attack, and becoming more vulnerable over time.

Assuming that we start in an uncompromised state, the initial conditions for these equations are $x_0(0) = 0$ and $x_n(0) = f_n X_T$.

For computations, I set $N = 5$ and used the distribution $f_n$ shown in the lower panel of Figure 7.2.

### 7.3.2.  *Cyber assets return with minimum vulnerability, which increases over time*

In this case, all reset cyber assets return to the pool $x_1$ (vulnerability equal to 0.5 corresponds to $n = 1$ in Figure 7.3). What happens after that? We will assume that except for $n = 5$ vulnerability increases in time, in the sense that cyber assets move from vulnerability $v_n$ to vulnerability $v_{n+1}$ at rate $\eta x_n$, where $\eta$ is a parameter. Hence, the dynamics for cyber assets with minimum vulnerability are

$$\frac{dx_1}{dt} = -ax_1v_1I(t) - a_{co}x_1x_0 + b\left(X_T - \sum_{n=1}^{N} x_n - x_0\right) - \eta x_1 \quad (7.6)$$

Cyber assets with intermediate vulnerability ($n = 2, 3, 4$) increase as hardened cyber assets lose vulnerability and decline as they lose hardening. Hence

$$\frac{dx_n}{dt} = -ax_nv_nI(t) - a_{co}x_nx_0 + \eta x_{n-1} - \eta x_n, \quad \text{for } n = 2, 3, 4$$
$$(7.7)$$

The most vulnerable cyber assets ($n = 5$) increase as cyber assets with vulnerability $v_4$ lose their hardening so that

$$\frac{dx_5}{dt} = -ax_5v_5I(t) - a_{co}x_5x_0 + \eta x_4 \quad (7.8)$$

The dynamics of the compromised assets are still described by Eqn. (7.5).

## 7.4.    Results

We separate the results according to whether the co-compromise rate parameter $a_{co}$ is smaller or larger than the threshold for the persistence of compromise. Although there is now a distribution of vulnerability among cyber assets, the basic intuition from Chapter 2 still holds: When $a_{co}$ is less than the threshold, long after the pulse attack has ended almost all cyber assets will have recovered to the uncompromised state. When $a_{co}$ is greater than the threshold, we know that compromise persists in the quasi-steady state.

Each figure of the results has four panels: The upper left panel shows the trajectories of the pulse, uncompromised, and compromised cyber assets; the upper right panel shows the trajectories of the uncompromised cyber assets with different vulnerabilities; the lower left panel shows the trajectory of mean vulnerability; and the lower right panel shows the trajectory of performance. Because the upper left and lower right panels show results with all cyber assets combined, they are directly comparable to results in Chapter 2. The upper right and lower left panels are the new ones and inform us about the temporal behavior of vulnerability.

### 7.4.1.  $a_{co}$ *is less than the threshold for persistence of compromise*

In Figure 7.4, I show the dynamics of the system when reset cyber assets return with vulnerability proportional to the discrete distribution shown in the lower panel of Figure 7.2. The upper left and lower right panels are comforting and accord with the results in Chapter 2 and our intuition. After the pulse ends, compromise is extinguished and the system returns to essentially all uncompromised cyber assets (upper left panel). The result of attack is a decline in performance of the cyber system or the enabled physical system, followed by return to full performance (the lower right panel).

Note the dynamics of the distribution of vulnerability as cyber assets are reset (upper right panel). During the pulse attack, the numbers of all cyber assets decline, with the relative decline determined by the vulnerability. Long after the attack ends, the distribution of cyber assets has a different distribution of vulnerability than the initial distribution of vulnerability. In particular, there are more of the least vulnerable assets in the quasi-steady state than at the start of the attack. This happens because we are not forcing the number of cyber assets with vulnerability $v_n$ to be fixed. Rather, all assets go into the single compromised pool and from there to resetting, and the fraction of that pool being reset to least vulnerable does not depend upon the number of current least vulnerable assets (in biology, we would say that there is no frequency dependence).

The consequence of these dynamics is that during the attack mean vulnerability declines (because the more vulnerable assets are removed at higher rates than the less vulnerable ones). After the pulse attack ends, the mean vulnerability increases, but does not return to its initial state.

When cyber assets return hardened but then become more vulnerable over time (Figure 7.5) the upper left and lower right panels

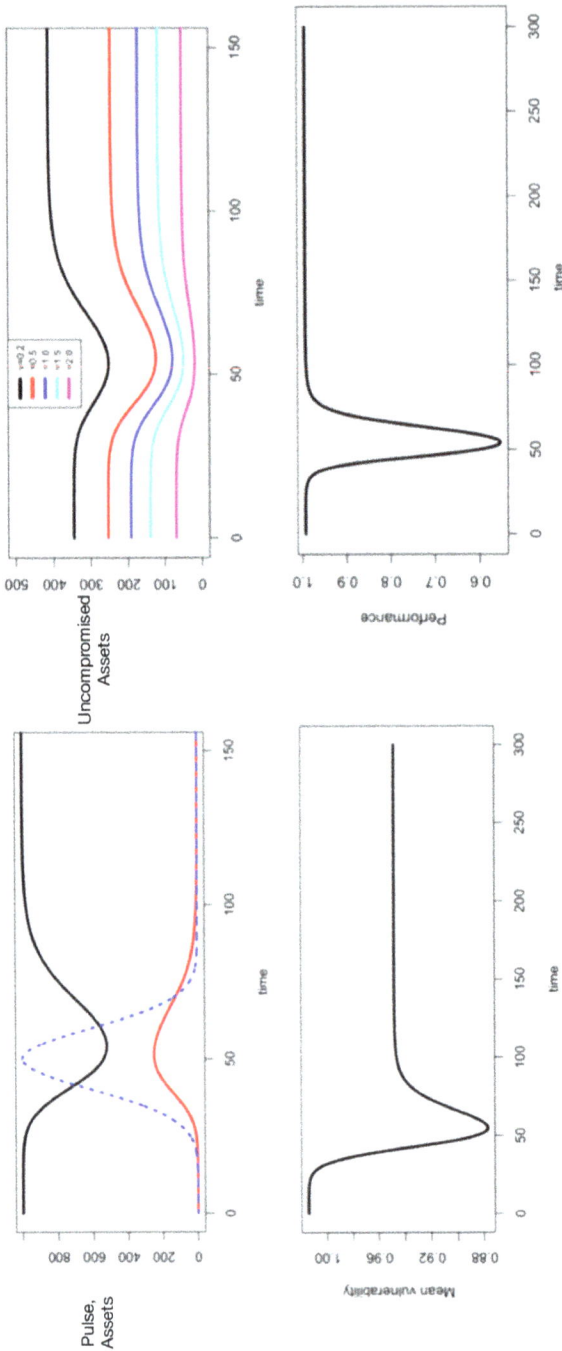

**Fig. 7.4.** The dynamics of the cyber assets and performance of the cyber or enabled physical system when $a_{co}$ is less than that the critical threshold for persistence of compromise and cyber assets are returned with vulnerability proportional to the initial distribution of vulnerability. The upper left panel shows the pulse attack (dotted blue line) and the total numbers of uncompromised and compromised cyber assets (black and red lines, respectively). The lower left panel shows mean vulnerability before, during, and after the pulse attack. The upper right panel shows the dynamics of the cyber assets separated according to vulnerability. The lower right panel shows the performance of the cyber system or enabled physical system. Note the different scale on the $x$-axes of the upper and lower panels.

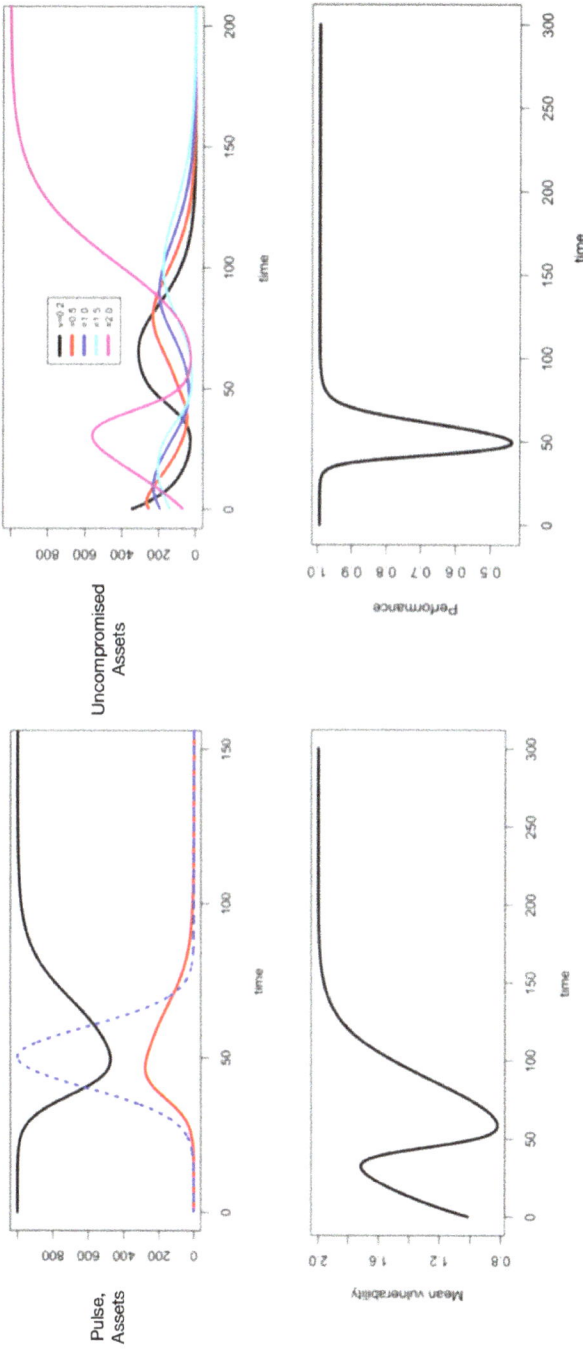

**Fig. 7.5.** The dynamics of the cyber assets and performance of the cyber or enabled physical system when $a_{co}$ is less than that the critical threshold for persistence of compromise in the steady state and cyber assets are returned at maximum hardening but then become more vulnerable as time progress. The upper left panel shows the pulse attack (dotted blue line) and the total numbers of uncompromised and compromised cyber assets (black and red lines, respectively). The upper right panel shows the dynamics of the cyber assets separated according to vulnerability. The lower left panel shows mean vulnerability before, during, and after the pulse attack. The lower right panel shows the performance of the cyber system or enabled physical system. Note the different scale on the $x$-axes of the upper and lower panels.

are similar to those in Figure 7.4, but the other two panels are very different. Because cyber assets return with minimum vulnerability and become more vulnerable over time, the number of cyber assets with vulnerability $v_1$ (the least vulnerable) peaks shortly after the attack ends; after that the numbers of cyber assets with vulnerability classes 1–4 decline as the most vulnerable class $x_5$ increases until essentially all cyber assets are at the highest vulnerability.

The consequence for mean vulnerability (Figure 7.5, lower left panel) is an increase and then decrease in mean vulnerability during the attack but then a rise towards maximum vulnerability as the quasi-steady state is approached.

### 7.4.2.    $a_{co}$ is greater than the threshold for persistence of compromise

When $a_{co}$ exceeds the threshold value, we know that compromise will persist in the quasi-steady state. As in the previous section, the upper left and lower right panels of Figures 7.6 and 7.7 match the results shown in Chapter 2. Thus, we should focus on the distribution of vulnerability over time.

When cyber assets are reset with vulnerability proportional to the initial distribution (Figure 7.6), in the quasi-steady state all cyber assets are at lower levels than their starting values (upper right panel), and in accord with our intuition the quasi-steady state levels are ranked according to vulnerability. In this case, mean vulnerability in the quasi-steady state returns to its initial value (lower left panel) because the numbers of all cyber assets are reduced from their initial values.

When cyber assets are reset with minimum vulnerability and then lose hardening as time progresses, we still have a peak (Figure 7.7, upper right panel). However, in the quasi-steady state least vulnerable cyber assets predominate because co-compromise is continually removing cyber assets that are more vulnerable, which are then reset and returned at minimum vulnerability. The consequence (Figure 7.7, lower left panel) is a rise in mean vulnerability as the attack starts and more vulnerable cyber assets are removed, a minimum of mean vulnerability at the peak of the attack, and then increase to a quasi-steady state value of mean vulnerability between them.

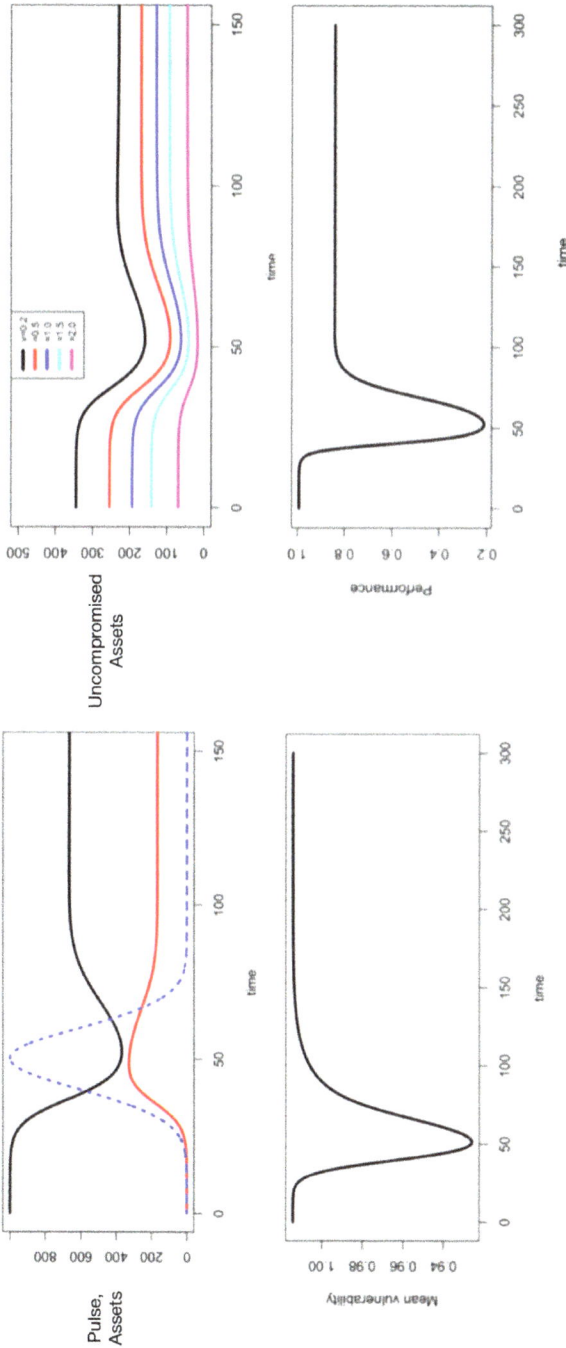

**Fig. 7.6.** The dynamics of the cyber system and performance of the cyber or enabled physical system when $a_{co}$ is greater than the threshold for persistence of compromise in the steady state and cyber assets are returned with vulnerability proportional to the initial distribution of vulnerability. The upper left panel shows the pulse attack (dotted blue line) and the total numbers of uncompromised and compromised cyber assets (black and red lines, respectively). The upper right panel shows the dynamics of the cyber assets separated according to vulnerability. The lower left panel shows mean vulnerability before, during, and after the pulse attack. The lower right panel shows the performance of the cyber system or enabled physical system. Note the different scale on the $x$-axes of the upper and lower panels.

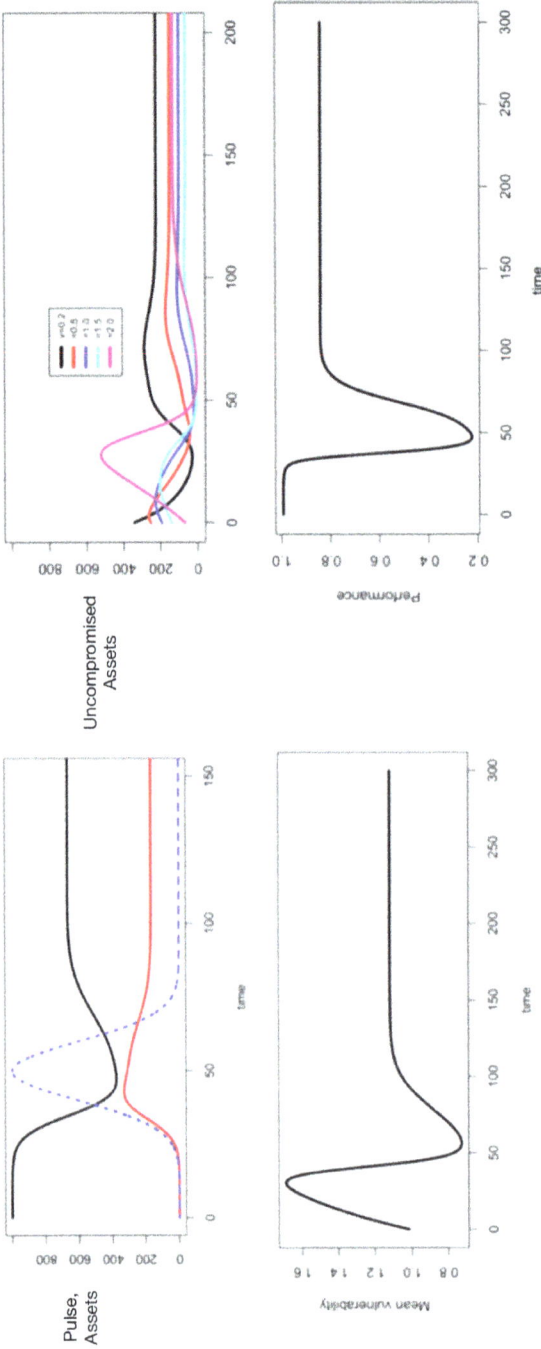

**Fig. 7.7.** The dynamics of the cyber system and performance of the cyber or enabled physical system when $a_{co}$ is greater than the threshold and cyber assets are returned at maximum hardening but then become more vulnerable as time progress. The upper left panel shows the pulse attack (dotted blue line) and the total numbers of uncompromised and compromised cyber assets (black and red lines, respectively). The upper right panel shows the dynamics of the cyber assets separated according to vulnerability. The lower left panel shows mean vulnerability before, during, and after the pulse attack. The lower right panel shows the performance of the cyber system or enabled physical system. Note the different scale on the $x$-axes of the upper and lower panels.

## 7.5. A Continuous Distribution of Vulnerability and an Open Question*

One could approximate the continuous distribution for vulnerability by taking $N$ in Eqn. (7.2) pretty large and setting the upper value of vulnerability to a value where the continuous distribution is close to 0. For example, suppose we set maximum vulnerability $v_{\max} = 5$ and let $N = 100$, with the $n$th vulnerability class $v_n = 0.05 + 4.95\frac{n-1}{N-1}$, so that $v_1 = 0.05$ and $v_N = 5.00$ and each increment in vulnerability is 0.05. We would then have 100 ordinary differential equations characterizing the dynamics of cyber assets with different vulnerability but they would be no more difficult to solve than what we have done (at least in principle).

Dwyer *et al.* (2000, pp. 116–117) showed how one can incorporate a continuous distribution of vulnerability into the equations for the SIR model of disease and thus understand how the distribution of vulnerability changes over time. To apply these ideas to the dynamics of cyber systems without wanting to assume perfect hardening of cyber assets after restoration requires extending the methods of Dwyer to the SIRS model of disease. In the context of cyber systems, it raises the question of how the continuous distribution of vulnerability changes when cyber assets are restored to the uncompromised state. This is an interesting, open but difficult question.

## 7.6. Summary of Major Insights

- The rate of compromise can be modified to include vulnerability to attack by assuming that cyber assets with different levels of vulnerability are compromised at rates proportional to their vulnerability. The gamma density pinned down at 0 is a flexible means of capturing a distribution of vulnerability to compromise.
- The gamma density can be discretized into $N$ values of vulnerability and the PAM expanded to include $N$ equations for the dynamics of cyber assets with different vulnerabilities.
- When vulnerability has a discrete distribution, key decisions have to be made about (i) how co-compromise occurs and (ii) when cyber assets are reset, how the vulnerability of reset assets is determined. In this chapter, we assumed that there is a single

pool of compromised assets and that co-compromise occurs at rate determined only by the rate of co-compromise, consistent with the assumption that the mechanism of external compromise and internal co-compromise are different. We investigated two choices for the vulnerability of reset cyber assets. In the first choice, reset cyber assets had vulnerability determined by the distribution of vulnerability at the time of the pulse attack. In the second choice, reset cyber assets return to the uncompromised pool with minimum vulnerability and then become more vulnerable as time goes on.

- When the co-compromise rate parameter is lower than the threshold for the persistence of compromise and cyber assets are returned proportional to the initial distribution of vulnerability, the overall number of uncompromised cyber assets and performance decline during the pulse attack but after the attack ends both return to the their values before the pulse attack. The numbers of cyber assets with different vulnerability decline during the attack and increase following the attack. However, because more vulnerable cyber assets decline at higher rates than less vulnerable assets, the distribution of vulnerability after the attack is different than before the attack and mean vulnerability may change from its value before the attack.

- When the co-compromise rate parameter is lower than the threshold for the persistence of compromise and cyber assets are returned with minimum vulnerability but then become more vulnerable as time progresses, during the pulse attack the overall numbers of uncompromised cyber assets and performance decline; after the attack ends both return to the their values before the pulse attack. In this case, there are transients in the number of cyber assets with different vulnerability, but ultimately all cyber assets have maximum vulnerability (unless other action is taken).

- When the co-compromise rate parameter is greater than the threshold for persistence of compromise and cyber assets are returned proportional to the initial distribution of vulnerability, during the pulse attack the overall numbers of uncompromised cyber assets and performance decline. After the attack ends compromise persists in the cyber system so that performance is permanently degraded. The numbers of cyber assets with different vulnerabilities decline during the attack and increase following the attack but not to their original levels. However, because more vulnerable cyber assets decline at higher rates than less vulnerable ones, the distribution of vulnerability after the attack is different than before the attack.

- When the co-compromise rate parameter is greater than the threshold for the persistence of compromise and cyber assets are returned with minimum vulnerability but then become more vulnerable as time progresses, during the pulse attack the overall numbers of uncompromised assets and performance decline. After the attack ends compromise persists in the cyber system so that performance is permanently degraded. The numbers of cyber assets with different vulnerabilities increase following the attack but not to their original numbers. Because more vulnerable cyber assets decline at higher rates than less vulnerable ones, the mixture of vulnerability after the attack is different than before the attack.

# Chapter 8

# Bon Voyage: Future Directions

*The story of the War of Atonement [the October 1973 Yom Kippur War] demonstrates the extent to which the human element is the key to the outcome of war... Even in the era of technology, the human element still stands at the centre of the picture*

– Herzog (2003, p. xvi)

I hope that you have enjoyed our exploration of how mathematical models from the population biology of disease can inform, facilitate, and enhance cyber security. One pleasure for me has been to see how relatively simple mathematics used in mature ways can lead to sophisticated understanding.[1]

There is much more to do! In this chapter, I will discuss three directions that I believe are fruitful. The first direction is explicit modeling of the cyber system of the enabled physical system. In this case, instead of a generic performance function we will focus on the electric grid. This involves a modest extension of the PAM.

The second direction is the inclusion of human factors, particularly operator attention, in the models for the dynamics of compromise and resetting. This will require the development of new models

---

[1]For example, we have not used game theory (e.g. Maynard Smith 1982, Axelrod 1984/2006, Axelrod and Iliev 2014, McNamara and Leimar 2020, Aurell *et al.* 2022) to treat adversarial strategies, dynamic programming (Alpcan and Basar 2011, Mangel and Clark 1988, Clark and Mangel 2000, Mangel 2015) to treat optimization, or Bayesian methods (Koch, 2007, Hobbs and Hooten 2015, McElreath 2020) to treat the uncertainty that is common in cyber system operations.

that capture the dynamics of human attention to compromise that are then linked to the PAM or FMSCO.

The third extension involves moving beyond dyadic interactions to multiple cyber actors. In this case, I suggest that we think about how cooperative **Multilateral Cyber Security Agreements** can be modeled. To help focus ideas, I will also be specific about the performance function. There is a large biological literature on cooperative defense and cooperative foraging and I will point towards some of it.

You can think of this send off chapter as a collection of three **Potential projects**.

## 8.1.    Modeling the Cyber System of the Enabled Physical System

Using the generic sigmoidal performance function for the enabled physical system allowed us to develop general principles. In specific applications, however, we require a model of the enabled physical system. For example, Mangel and McEver (2021) used a nonlinear oscillator model of the electric grid (Filatrella *et al.* 2008) to explore how compromise of smart meters can lead to instability in the electric grid. But even there, the cyber components of the grid were not modeled.

In general, to enable a physical system by a cyber system requires cyber components in the enabled physical system. The closest we came to modeling the cyber components of the enabled physical system was Eqn. (1.5) and Figure 1.7 when we considered that performance of the enabled physical system might not be a sigmoidal function of the number of uncompromised cyber components, but a sigmoidal function of the performance of the cyber system itself.

For definiteness, to help us think about the cyber components that link the focal cyber system and the enabled physical system let us focus on an electric grid providing power to consumers (National Research Council 2013). Libicki (2016 pp. 187–188) noted

> Power grids are an obvious target for disruption. In 2013, the director of national intelligence deemed a large-scale cyberat-tack on the nation's critical infrastructure (of which the electric grid is the most prominent part) the greatest short-term threat to the nation's security. One analyst has calculated that

even a temporary shutdown of the power grid could cost the United States 700 billion dollars (that is, more than the defense budget)...

...It might seem odd that a power grid that worked well before the Internet arrived could be so vulnerable, but over the last several decades, power companies have concluded that it is far cheaper to have each of thousands of power stations, transmission lines, and distribution centers answer to a central information service than to send people on trucks to check on equipment every time something does or might happen. Therein lies a problem if hackers can work their way into the electric grid system and send it the kind of commands that tamper with, say, voltage levels. Overloaded circuits, for instance, can create a cascading failure that takes down entire systems.

Let us thus envision a situation as in Figure 8.1, where the left cyber system corresponds to the utility company and the right cyber system corresponds to the generating plant:

- The cyber system of the utility company receives input from consumers (here shown as a partially blue arrow when the input is uncompromised and a partially red arrow when the input is compromised) and external attack (solid red arrow).
- The cyber components of the utility company follow dynamics similar to the PAM, and send signals that are uncompromised (green) or compromised (red) to the cyber system of the generating plant.
- The cyber components of the generating plant also follow dynamics similar to the PAM, and send signals that are uncompromised (green) or compromised (red) to the generator, and these signals determine the behavior of the generator.

I hope that it is clear why we did not begin the book here. The additional cyber system doubles the number of equations and parameters describing the dynamics of the cyber system. In addition, we need to model the dynamics of the generator responding to the cyber system controlling it.

> **Potential project\***: Convert these verbal and visual models to a set of equations for the two cyber systems. Then explore linking the nonlinear oscillator model for a generator (Filatrella *et al.* 2008) to the two sets of PAM-like equations and explore the consequences.

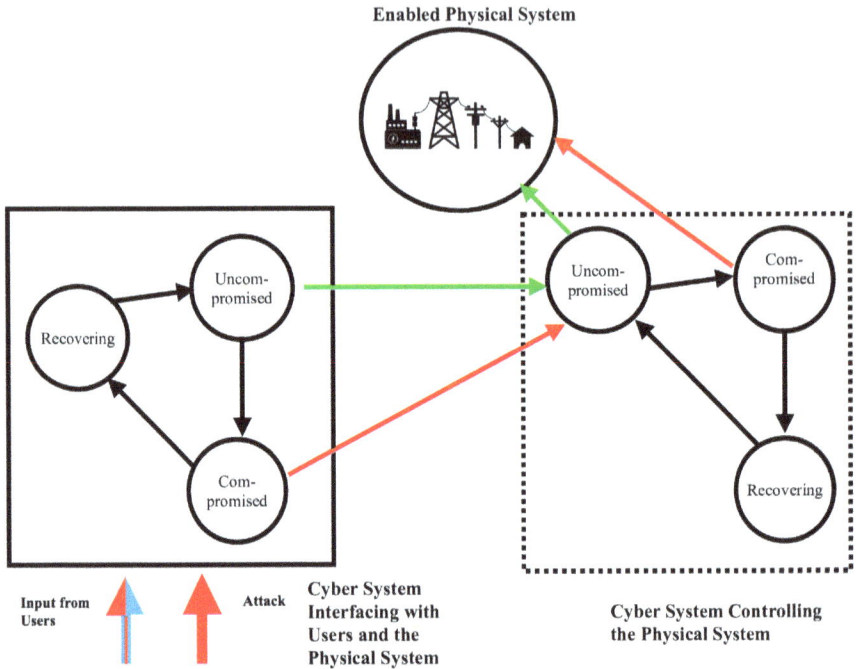

**Fig. 8.1.** A model that explicitly accounts for the cyber component of the electric grid. There are now two cyber systems. The first one (left panel) interacts with users of the electric grid, receiving input from the users and possibly being subject to attack (see Mangel and McEver 2021 for an example). This cyber system then interacts with the cyber system that controls electric generator and associated transmission features (right panel), and sends signals to the generator.

## 8.2.  Human Factors, Particularly Operator Attention

Cyber security is not just a technical matter (Danzig 2014, Meyer 2024). Indeed, Lindsay and Gartzke (2018, p. 199) write "No discussion of coercion would be complete without some attention to the psychological dimension". Libicki *et al.* (2015) asked about 20 cyber security officers what they would do if provided more money for cyber security, and a majority of them gave solutions that were human-centric.

Libicki (2016, p. 26, 43ff) notes the important role of human factors in the redundancy component of the resilience stack, writing that redundancy "gets easier if human judgment can be applied – that is, if these computers are information gatherers for human decision makers". In fact, Libicki asks the rhetorical question "Is cybersecurity a technology or a people problem?", giving the answer yes, and considers (Libicki 2016, p. 65) that "The most reliable measure (and counter-countermeasure) is a smart analyst who can use human experience and intuition to detect when malware rather than legitimate traffic is touching the sensors and infer tools from instances of sensor disturbance. But while software scales well (in that it can be used anywhere), humans do not; furthermore, the ones that can do these jobs are expensive. Not surprisingly, the optimal man-machine ratios will shift over time. If serious infections are rare (for example, if the average organization goes years between attacks), scalability is less important, and throwing smart people at the infection in order to extract tool signatures can be justified."

The Aurora generator test (Zetter 2014) provides a good example of human factors. In that case operators who monitored the grid for anomalies "weren't told of the attack before it occurred never noticed anything amiss on their monitors. The safety system that was designed to ride out little spikes and valleys that normally occurred on the grid also never registered the destructive interruption. 'We could do the attack, essentially open and close a breaker so quickly that the safety systems didn't see it,' said Perry Pederson, who headed DHS's control-system security program at the time and oversaw the test" (Zetter 2014, p. 164).

Our discussion in Chapter 5 about the probability of a kinetic response has an implicit human factor in it. Indeed, Libicki (2016, p. 225) writes "The attacker always has to factor in some likelihood that an attack will engender a response. The questions are: how likely and how bad? A lot depends on the likelihood in the mind of the attacker that the target of the contemplated attack detects the cyberattack (something that is not obvious in a corruption attack), identifies the attacker correctly with an actionable level of confidence, judges that these attacks crossed some threshold, decides to strike back even in the face of counter-threats that the attacker can make to ward off retaliation, can strike back, and thereby cause the attacker real pain."

Thompson (2022, p. 10) writes that "... the human brain is also an inseparable and vital part of modern decision-making systems. Going further, this is a partnership: the human brain is responsible for constructing models; models provide quantitative and qualitative insights; the brain can integrate these with other, non-modeled insights; and the upshot is a system that can be better than either brain or model acting alone." Consideration of human factors could include:

- Assuming that users of cyber systems will have both security training, and may not learn much or may forget what they have learned.
- Recognizing that nodes of the cyber system that interface with the external world are another weak link in the chain of security, and human attention to those nodes may have big payoffs.
- Recognizing that unlike machines which are usually limited in the kinds of inputs that they receive, humans have many inputs working simultaneously, sometimes providing contradictory information, and humans are good are noticing anomalies. Libicki (2016, p. 44) gives the following example concerning a USB-based attack: "A hacker could embed keyboard-like programs into a USB device that the users, having no ostensible reason to believe it is not a memory device, stick into a computer. Unbeknownst to the user, the USB device could be entering malicious commands via simulated keystroke into the computer (which a computer will accept even if not a response to a query). One of these commands could have the computer similarly infect all USB devices consequently linked to the computer, which then infect other computers they are inserted into. This is a tricky but not unstoppable hack. One approach is to have all USB devices signed for what they are and what they are allowed to do; this would require coordination among a great number of vendors (including some who have since left the business and whose devices thus would no longer work once signatures became mandatory). *A far simpler approach is for a computer's operating system to announce to the user what device it is (and hence what privileges it has been accorded). A person inserting a USB memory stick, but programmed to act as a keyboard, into a machine would be greeted by a message that a keyboard had been inserted. Many users would then balk and unplug the stick. Not all users would, but any attack that requires spreading the USB throughout the organization without anyone noticing it would have slim chances of success*" (italics added).

Recognizing an anomalous situation may be difficult for either a person or a machine. For example, one of the analysts trying to understand Stuxnet "believed the attackers deliberately used weak encryption and a standard protocol to communicate with the servers because they wanted the data traveling between infected machines and the servers to resemble normal communication without attracting unusual attention. And since communication with the servers was minimal – the malware transmitted only limited information about each infected machine – the attackers didn't need more advanced encryption to hide it" (Zetter 2014, p. 63).

This very brief discussion of an important topic suggests that we need to modify Figure 8.1 to include operator attention in both cyber systems, as in Figure 8.2.

Adding operator attention requires an additional set of equations, and the associated assumptions. Perhaps the best way to develop them is to personally experience the cyber operational that you

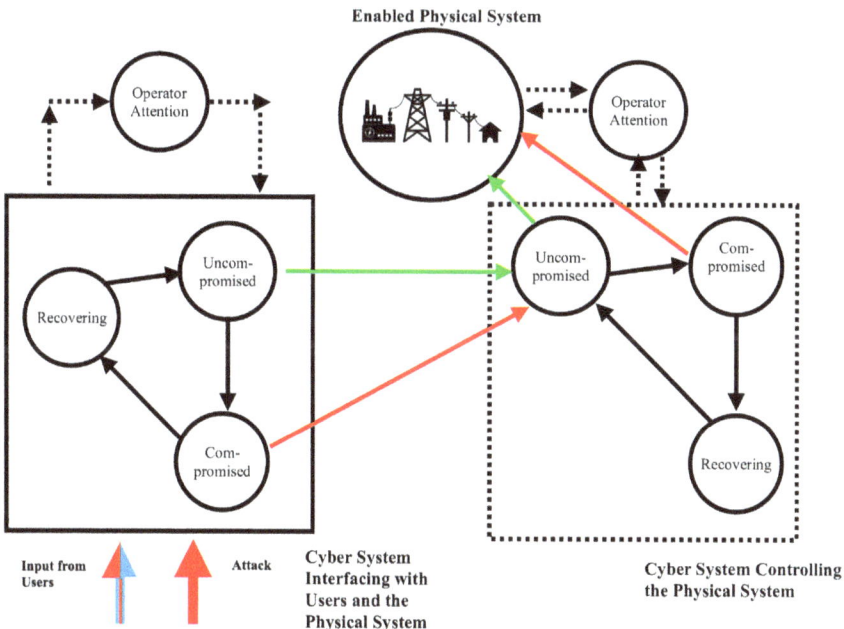

**Fig. 8.2.** A model that explicitly considers the cyber components that enable the physical system when operator attention is included, with the electric grid as a motivation. See text for explanation.

are trying to model (Mangel 1982). For now, here is a potential project.

> **Potential project**: Focus on the PAM and incorporate operator attention via an explicit model for the detection of compromise, allowing detection to depend upon any combination of the number of compromised cyber assets, the rate at which cyber assets are compromised, or reduction in performance. When compromise is detected, allow the operator to introduce defensive counter-measures (*sensu* Mangel and McEver 2021) by choosing to increase either $r_x$ or $b$, but with a reduction in performance. Code the model and explore its consequences.

The classic works of Janis and Mann (1976, 1977) and Sage (1981) remain some of my favorite pieces on operator attention. Starting points for more recent literature include: Pirolli (2009) who applies ideas from biological foraging theory to the adaptive interaction between humans and machines searching for information; Kahneman (2012) who provides a great place to start learning about human thinking; McDermott (2010) who discusses decision making under uncertainty with cyber issues in mind; and Johnson *et al.* (2017), who offer a model of human visual attention and workload when controlling robots.

## 8.3.   Beyond Dyadic Interactions: Towards Multilateral Cyber Security Agreements

Until now, we focused on dyadic interactions and had very general performance functions in both the PAM and the FMSCO. An important future direction is to increase the number of potential cyber actors (now assumed to be nations) and be very specific about performance. Since this topic is very different from the PAM and FMSCO and is important, I provide detailed steps on how you can get started.

Valeriano and Maness (2015, p. 201) offer guidelines for justice and behavior in cyber space and write: "There is a need to move toward creating a global monitoring system to share in collective defense and punishment when cyber violations occur." In March 2023, the White House released a document describing a national cyber security strategy (White House 2023) based on five pillars. One pillar is "Forge International Partnerships to Pursue Shared Goals". In this chapter, we will develop ideas that provide a conceptual framework for collective action described by the guideline of Valeriano and Maness (2015) and the pillar of the White House document. In doing this, we will think about joint operations in cyberspace (White 2020) and cumulative deterrence (Tor 2017).

I want us to envision and think about how to model a **Multilateral Cyber Security Agreement (MCSA)** that captures the interests of many member nations of the international community simultaneously (Sandler 2004, Benvenisti and Nolte 2018, Zabierek *et al.* 2021). Transparency (Chayes and Chayes 1995) is important in a MCSA and mathematical modeling helps in this case because it requires a precise statement of ideas. There is already a tradition in international law of using mathematical models and statistical models as proof of causality (e.g. in the Gulf War Reparations Claims (Sulyok 2021, p. 308)). We can also aim to help lawyers and policy makers become better consumers of models (Sulyok 2021, Thompson 2022).

There are analogies in the population biology of disease. For example, the Global Influenza Surveillance and Response Network (the "flu network" (Kapczynski 2016, Stein 2020)), established in 1952, is a multilateral global heath cooperative and played a key role in the early recognition of the COVID-19 pandemic. The successful global eradication of smallpox (Fenner 1982, Henderson 1987) was based on a collaborative effort by nations of the world. Although not yet completed the Global Polio Eradication Initiative launched in 1988 is a multilateral effort (Aylward and Tangermann 2011, Chumakov *et al.* 2021).

### 8.3.1. *Some ideas from international law*

Even though progress has been made (Schmitt and Vihul 2017) many legal aspects of cyber conflict remain blurry (Singer and Friedman 2014, p. 122 ff). For example, it is already clear that the equivalent of Article 5 of the NATO Treaty, which asserts that attack on one member shall be considered an attack on all members, is not directly applicable to cyber attack (or if it is, then it applies in a much more nuanced way). Part of this issue is one of attribution: when a cyber attack occurs, to respond one needs to answer "who did it?" with sufficient confidence. Such confidence may often be lacking (Rid and Buchanan 2015). For this reason, we focus only on cooperative cyber defense and ignore questions about cyber counter-attack.

We will assume that nations will join a MCSA if they do better by joining, accounting for relevant costs, than by going alone in defense. In such a case, coordination in global cyberspace (Trachtman 2013, p. 112) can be formulated as the stag hunt game (Binmore 1994, p. 120ff, Skyrms 2004). The fundamental idea is that in a population of hunters, individuals are able to capture rabbits by themselves but require the cooperation of and trust in other individuals to capture a stag. Thus participation in the stag hunt requires that the individual's expected share of the stag is bigger than the expected catch of rabbits when working alone.

It is beyond the scope of this chapter to try to review the history of environmental treaties. Some starting points are Chayes and Chayes (1996), Barrett (2003), Andresen (2014), Gupta (2014), and Brunnée (2018). For example, a variety of Multilateral Environmental Agreements (MEAs) incorporate scientific analysis as foundational. These include the International Convention for the Regulation of Whaling (1947), the Agreement on the Conservation of Polar Bears (1973), the Convention on the Conservation of Migratory Species of Wild Animals (1979), the Convention for the Conservation of Antarctic Marine Living Resources (1980), and the The Montreal Protocol on Substances that Deplete the Ozone Layer (1987). The current text for any of these agreements can be quickly found by an internet search.

Sulyok (2021, p. 352ff) calls for an approach incorporating science in law that is "hybrid", in the sense of neither purely legal/policy nor purely modeling, arguing that such a hybrid approach provides a firm framework for framing legal questions that have considerable scientific and/or modeling components. This approach does not require that decision or policy makers become modelers, but does require that modelers describe their assumptions and results in a way that the key ideas are accessible to decision or policy makers.

### 8.3.2. *An overview of what needs to be done*

We begin with a a specific metric of performance. For definiteness, I chose the connection with **Broad Band Penetration (BBP)** and **Gross Domestic Product (GDP)** for countries in the **Organization for Economic Cooperation and Development (OECD)**. We will see that these can be related using a simple linear regression.

Cyber attack will reduce the available BBP of a nation, which then reduces GDP according to the linear regression characterizing performance. We then ask (i) if a nation can allocate some of its BBP to defense (as an organism can allocate some of its energetic reserves to defense against pathogens), how does performance (GDP) vary with allocation to defense and (ii) if there is the possibility of joining a MCSA, when is it advantageous for a nation to join and how does that depend on the behavior of other nations?

We first compute the defense that maximizes the national value, defined as the GDP obtained when some of BBP is allocated to defense rather than towards GDP. We will see that there is an optimal level of BBP to dedicate to defense, and that the peak (for the parameters we use) is broad.

Multilateral defense depends on the behavior of nations. A starting assumption is that each nation honors its multilateral commitment with certainty. We can then compare GDP when a nation goes it alone or to the GDP if it joins the MCSA. Alternatively, a nation may not fully trust the other nations to honor their commitment so that we require a **National Trust Factor (NTF)** that is the probability characterizing a nation's belief that other members of the MCSA will honor their commitments to monitoring and defense. The NTF emerges through repeated interactions between nations (e.g. Axelrod 1984/1986) and characterizes how a nation views the the propensity of other nations to comply with previous agreements, which itself is determined by a variety of factors such as efficiency of the agreement, the interests of other nations, and international norms (Chayes and Chayes 1995). It is, at least in principle, something that can be determined from experience and government policy.

Since the NTF is a probability, we can use a simulation to create multiple realizations of national values obtained by joining a MCSA or not. This is different from the simulations employed in Chapter 4. There, we drew random numbers to determine if a change in the state variables occurred in the next bit of time, and then which change occurred given that there was a change. We used the deterministic dynamics of the PAM to guide setting up the probabilities of change.

In this case, we repeatedly draw random variables to determine a nation's BBP and its NTF, and use these to compute the national values that determine whether it joins the MCSA or not. Thus, over multiple realizations of the simulation, we capture the probability or frequency with which a given nation joins the MCSA.

Once this simulation for a hypothetical set of nations is completed, we can apply it to assembling the MCSA for OECD countries. In this case, we would use the BBP for each OECD country, rather than simulating value of BBP, but still draw NTFs randomly, with the goal of computing the fraction of nations joining a MCSC that is initiated by a fraction of the OED nations.

### 8.3.3.   *Some details to get you started*

I just outlined a challenging plan of research, so now give a few details to get you started.

#### 8.3.3.1.   *Broad band penetration and gross domestic product in OECD countries*

In Figure 8.3, I show the relationship between GDP and BBP in OECD countries. When fitting the line, I treated Ireland and Luxembourg as outliers, which left 35 data points, and henceforth refer to these as the OECD nations. We will use the fitted line as the performance function for assessing consequences of cyber attack and both unilateral and multilateral cooperative defense; it is $GDP = 3.84 + 1.17 \cdot BBP$.

To fit the data in Figure 8.3 to a **beta density** (Hilborn and Mangel 1997, Mangel 2006), we first map the data onto the interval [0,1] by introducing a variable $X$ representing a scaled version of the data, using a minimum value of BBP equal to 15.3 and a maximum value equal to 50.3 (i.e. 35 BBP units higher than the minimum), so that

$$X = \frac{BBP - 15.3}{35} \qquad (8.1)$$

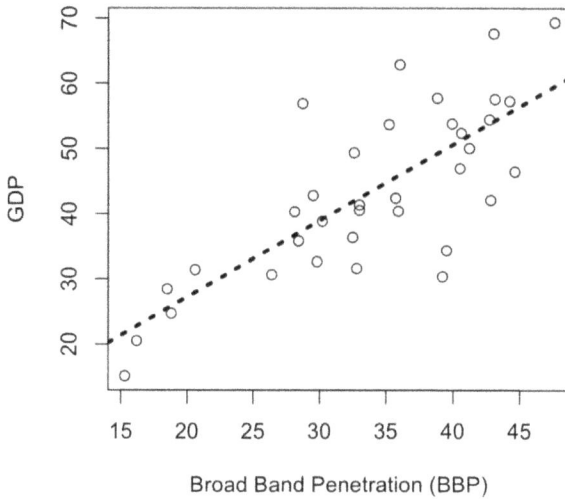

**Fig. 8.3.** The relationship between BBP, measured as total subscriptions per 100 inhabitants and per capita GDP, measured as 1000s of USD per person annually from OECD data (OECD 2017). For the regression, I considered Luxembourg and Ireland as outliers. The fitted line is $GDP = 3.84 + 1.17 \cdot BBP$.

Next we assume that $X$ has a beta density with parameters $\alpha$ and $\beta$

$$f(x|\alpha, \beta) = \frac{\Gamma(\alpha + \beta)}{\Gamma(\alpha)\Gamma(\beta)} x^{\alpha-1}(1 - x)^{\beta-1} \tag{8.2}$$

where $\Gamma(\cdot)$ is the gamma function from Chapter 7. As there, we can think of $\frac{\Gamma(\alpha+\beta)}{\Gamma(\alpha)\Gamma(\beta)}$ as a normalization constant.

If a random variable has density given by Eqn. (8.2), its mean and variance are $\frac{\alpha}{\alpha+\beta}$ and $\frac{\alpha\beta}{(\alpha+\beta)^2(\alpha+\beta+1)}$ respectively. If $m_x$ and $v_x$ are the mean and variance of a beta distributed random variable and $m_x(1 - m_x) > v_x$ then the method of moments estimates for the parameters are

$$\alpha = m_x \left( \frac{m_x(1 - m_x)}{v_x} - 1 \right)$$

$$\beta = (1 - m_x) \left( \frac{m_x(1 - m_x)}{v_x} - 1 \right) \tag{8.3}$$

This can be verified by substituting Eqns. (8.3) into the formulas for mean and variance below Eqn. (8.2) and seeing that they reduce to $m_x$ and $v_x$.

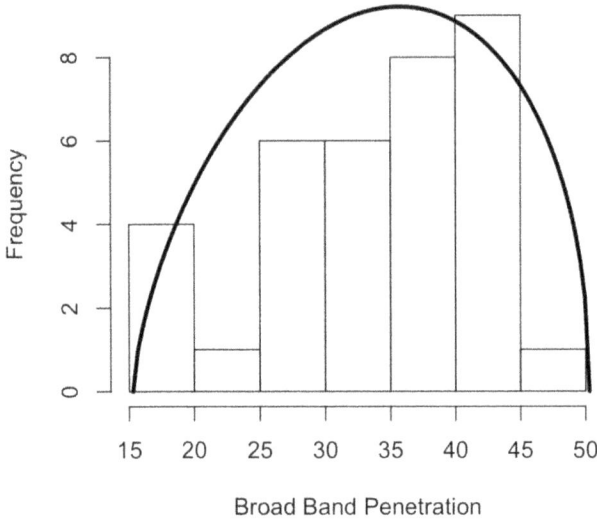

**Fig. 8.4.**    A histogram of the OECD data shown in Figure 8.3 and the beta density fitted to it by using the method of moments.

In Figure 8.4, I show a histogram of the OECD data and the fitted beta density.

### 8.3.3.2.    *Cyber attack and its consequences*

We let $\lambda(b)$ denote the annual number of cyber attacks on a nation with BBB $= b$, modeled as

$$\lambda(b) = \lambda_{\min} \left[ \frac{b}{b_{\min}} \right]^{\beta_a} \tag{8.4}$$

depending on parameters $\lambda_{\min}$ characterizing the minimum rate of attack, which occurs when BBP is the minimum value $b_{\min}$, and $\beta_a$ characterizing how the rate of attack varies as BBP exceeds the minimum value. Figure 8.5 shows this function for $\lambda_{\min} = 50, b_{\min} = 10$ $\beta_a = 1.35$, and set the maximum value of $b$ to 60.

We then assume that the loss of BBP due to attack is $[\lambda(b)l_a]^k$ where $l_a$ is the loss per attack and $k$ is a shape parameter. Setting $\tilde{l}_a(b) = \lambda(b)l_a$, if a country with BBP $= b$ executes no defense, its resulting BBP is $b - \tilde{l}_a(b)^k$. I illustrate this in Figure 8.6 with $l_a = 0.05$ and $k_a = 1.05$.

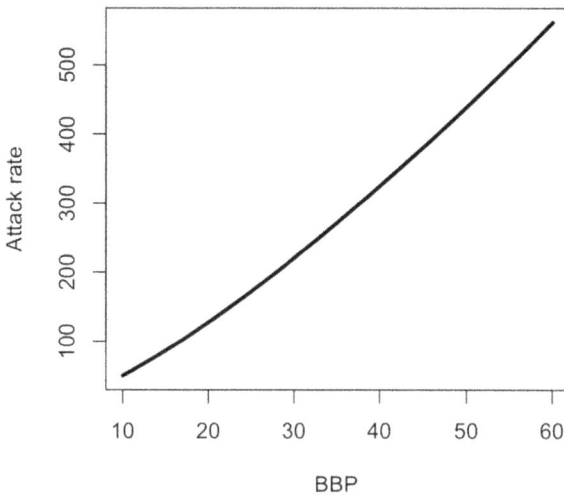

**Fig. 8.5.** The rate of cyber attack on nations as BBP varies from a minimum value of 10 to a maximum value of 60, based on $\lambda(b) = \lambda_{\min} \left[ \frac{b}{b_{\min}} \right]^{\beta_a}$ with parameters $\lambda_{\min} = 50, b_{\min} = 10$ and $\beta_a = 1.35$.

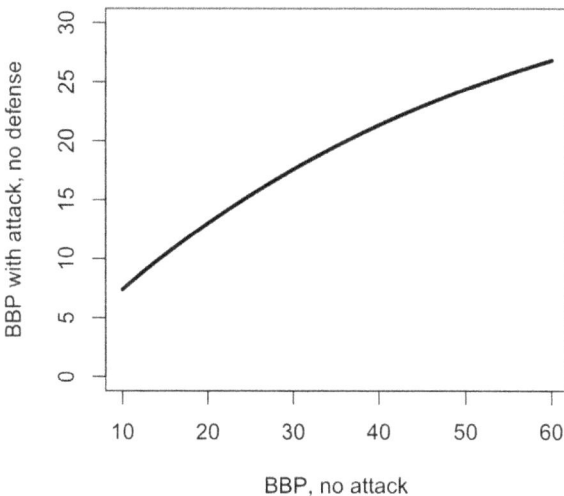

**Fig. 8.6.** The reduction in BBP due to cyber attack following Eqn. (8.1), loss per attack $l_a$, and overall reduction in BBP a power function of expected loss. For a nation with BBP $b$ in the absence of cyber attack, BBP after cyber attack, absent defense, is $b - [\lambda(b)l_a]^k = b - \tilde{l}_a(b)^k$.

### 8.3.3.3.  *Defense maximizing national value*

Libicki *et al.* (2015, p. 2) note that "the proper goal of a cybersecurity program (or policy) is to minimize the combined cost of expenditures on cybersecurity plus the expected costs arising from cyberattacks". We now ask what happens if a nation with BBP $b$ allocates some of its BBP, which we still call cyber assets, to defense, thus reducing BBP available for economic activity and in consequence GDP. To capture this idea, we assume when $z$ BBP cyber assets are allocated to defense, the fraction of attacks that are successful is $e^{-wz}$, where $w$ is a measure of the effectiveness of defense.

With this formulation, the net BBP for a nation with initial BBP $b$ that allocates $z$ assets to cyber defense is

$$b_{\text{net}}(b, z) = b - [\tilde{l}_a(b)e^{-wz}]^{k_a} - z \tag{8.5}$$

The total cost to a nation is the sum of BBP assets used in defense and those lost to attack, so that for a nation with BBP $b$ allocating $z$ assets to defense, the cost is

$$C(b, z) = z + [\tilde{l}_a(b)e^{-wz}]^{k_a} \tag{8.6}$$

Taking the derivative of $C(b, z)$ with respect to $z$ and setting it equal to 0 gives the cost minimizing defense

$$z^*(b) = \frac{1}{wk_a} \log\left[\tilde{l}_a(b)^{k_a}wk_a\right] \tag{8.7}$$

For computations, I used $w = 0.2$ and set the maximum possible level of defense to be $z_{\max} = 10$; for the other parameters in the model, this constraint was not binding. In Figure 8.7, I show the allocation to defense as an absolute function of BBP (left panel) and as a fraction of national BBP (right panel). We see that i) when a nation's BBP is small enough, the optimal allocation to defense is 0 and ii) otherwise the allocation to defense is an increasing function of BBP at a decreasing rate of increase so that the fraction of BBP dedicated to unilateral defense has a broad peak and then declines. Unilateral defense here is analogous to baseline defenses described by Buchanan (2016, p. 185).

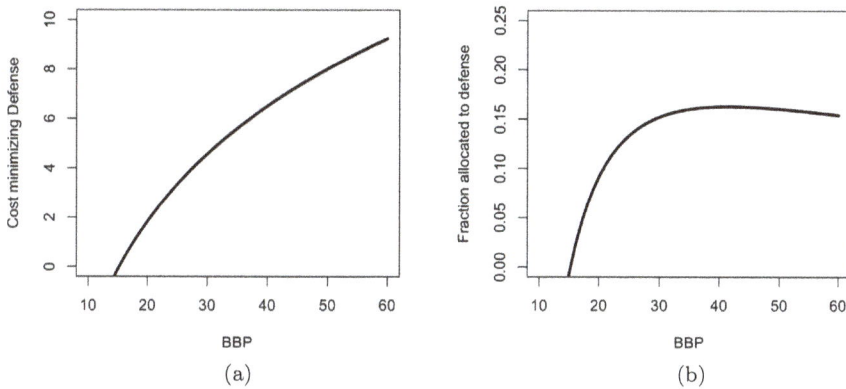

**Fig. 8.7.** The cost minimizing defense, Eqn. (8.7), as a function of BBP (left panel) and the fraction of BBP allocated to defense when the cost minimizing defense is used (right panel).

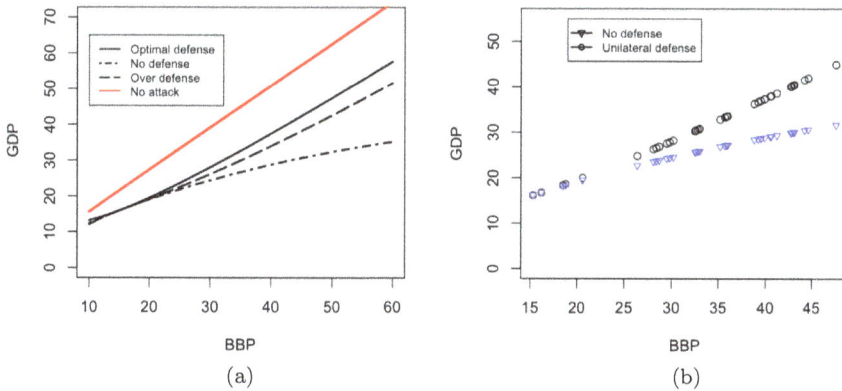

**Fig. 8.8.** Left panel: GDP if there were no cyber attacks at all (red line), cyber attack with no defense, cyber attack with optimal defense, and cyber attack with twice the optimal defense as a function of BBP. Right panel: GDP with no defense and unilateral optimal defense for the OECD countries in Figure 8.3.

Each choice of defensive allocation $z$ in Eqn. (8.7) leads to a net level of BBP

$$b_{\text{net}}(b, z^*(b)) = b - z^*(b) - [\tilde{l}_a(b)e^{-\omega z^*(b)}]^{k_a} \qquad (8.8)$$

which then determines GDP based on the regression line in Figure 8.3. In Figure 8.8, I show GDP if there were no cyber attacks at all (as a reference case), with no defense, with optimal defense and twice the optimal defense (left panel). We conclude that, given the other parameters in the model, it is better to over-defend (twice the optimal level of defense) than to not defend at all. In the right panel of Figure 8.8, I show GDP if there is no cyber defense and if there is optimal cyber defense for the OECD countries.

The results in Figure 8.8b show that, given the other parameters in the model, there is an advantage of unilateral cyber defense (Eqn. (8.7)) over a wide range of BBP for the OECD countries. We now turn to multilateral defense.

---

**Potential project**: Suppose that GDP was a nonlinear function of BBP, so that a country with BBP $b$ has GDP $g(b)$. In that case, $GDP(b, z) = g(b_{\text{net}}(b, z))$. What can be said about the GDP maximizing defense?

---

### 8.3.4.   *Modeling the MCSA*

These preliminaries take us to the point where we can think about the MCSA:

(1) Imagine $N$ nations each of which chooses to participate in the MCSA or not, with the BBP of nation $n$ denoted by $b_n$. Each nation commits a fraction $\delta_{\text{MCSA}}$ of its defensive BBP $z^*(b_n)$ to multilateral cyber defense. $\delta_{\text{MCSA}}$ is something to be negotiated between the nations participating in the MCSA.

(2) Thus, the defensive BBP from the MCSA is the sum of the allocations from each nation, that is the sum of $\delta_{MSCA} z^*(b_n)$ for $n = 1, 2, \ldots, N$. We are now faced with a modeling decision about how the total defensive BBP in the MCSA are allocated between the participating nations. Clearly all the nations cannot obtain all of the defensive BBP in the MCSA, so a starting point is to assume that each of the nations participating in the MCSA receives an equal fraction of the joint defensive BBP. A nation's total defensive BBP is then the sum of $(1 - \delta_{\text{MCSA}}) z^*(b_n)$ (the BBP it is not sharing) and the share of the defensive BBP in the MCSA.

> **Potential project**: Think about other ways that the total defensive BBP in the MCSA $z_{\text{MCSA}} = \delta_{\text{MSCA}} \sum_{n=1}^{N} z^*(b_n)$ could be allocated across the nations and how the enhanced defense could be defined.

(3) When nations are guaranteed to honor the MCSA, given the total defensive BBP each nation can then compute GDP with unilateral cyber defense and multilateral cyber defense for the OECD countries and by comparing them decide whether to participate in the MCSA or not.

(4) When nations are not guaranteed to honor the MCSA, the simplest case is that the NTF is constant across all nations. Then an individual nation would simply reduce the defensive BBP from the MCSA by multiplying by the NTF and proceed as just described.

(5) The situation is more complicated when each nation has its own NTF, but readily treated by simulation methods. We would start each run of the simulation of the behavior of $N$ nations by drawing the BBP and NTF of each nation. Values of BBP could be drawn from the beta density associated with Figure 8.4 (i.e. Eqn. (8.2)). To give $N$ nation its own NTF, $\phi_n$, I would also use a beta distribution with specified mean and variance. For example, in Figure 8.9 I show two beta distributions with the common mean to 0.7, but variance either 0.02 or 0.2.

(6) In each run of the simulation, we cycle over nations and for each nation compute its GDP if it goes alone and its total defense and then GDP if it participates in the MCSA. We then would allow each nation to join the MCSA or not. By repeated running this simulation, with randomly drawn BBP and NTF for each nation, we will generate a distribution for the number of nations participating in the MCSA.

When this is done, we are ready for the prize: Assembling a MCSA for the OECD nations. To do so, imagine that a small number $N_{\text{init}}$ of the OECD nations have decided that they want to establish a MCSA. We no longer would need to choose the BBP randomly, because they are known for the OECD nations, but still need to choose the NTFs from a distribution. It makes sense to me that the

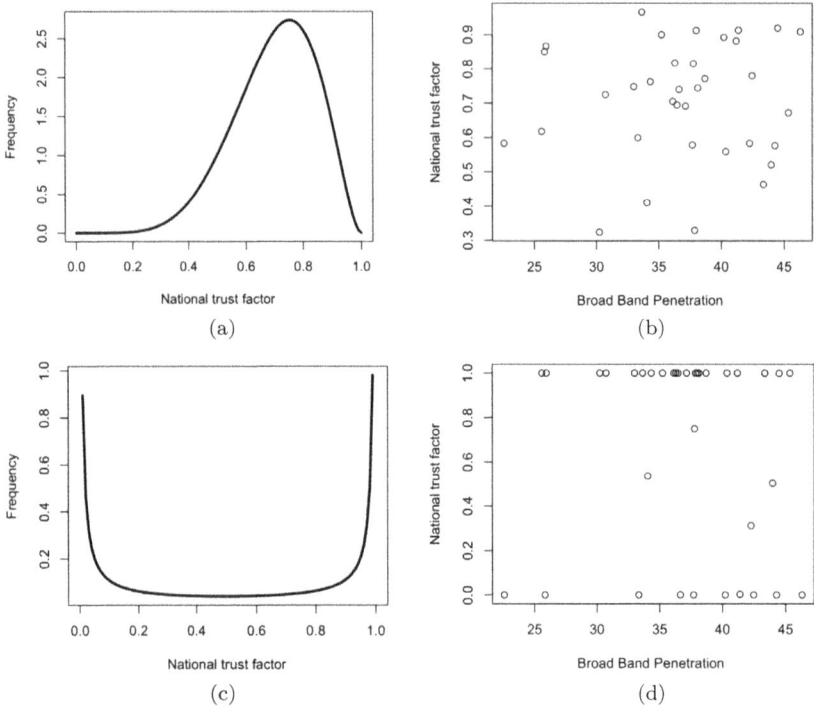

**Fig. 8.9.**    Left panels: Two choices for the beta density of NTFs. Both distributions have the same mean (0.7) but the variance is 0.02 in the upper panel and 0.2 in the lower panel. We might describe the upper panel as "most nations trust each other somewhat" and the lower panel as "some nations are very trusting, while others are very untrusting". In the right panels, I show examples of simulated NTFs.

NTFs of the initializing nations should be at least the mean value or higher. After the MCSA is initialized by those first $N_{init}$ nations, we can ask sequentially if the remaining nations would choose to join the MCSA or not, by comparing GDP with unilateral defense and GDP with multilateral defense, and then making a rational choice (Hechter 1987, Guzman 2008) of joining the MCSA if GDP is larger when in the MCSA than GDP when the nation goes it alone.

Simulation has a role to play here too. Without specific knowledge of the policies of the nations, there is no *a priori* means for choosing the initial nations or the order in which nations choose to join the MCSA or not. To deal with this problem, we can randomly re-order the OECD nations in each run of the simulation (which can be done using the "Sample" package in R without replacement) and follow the procedure described in the previous paragraph. Each replicate of the simulation will generate the fraction of the nations joining the MSCA, so that after all of the replicates are run we have a distribution for the fraction of nations joining the MCSA.

## 8.4. Create Your Own Voyage

These examples (modeling the cyber system of the enabled physical system, modeling human factors in the PAM and the FMSCO, and modeling a MCSA) are exactly that – examples of future directions. Of course, I hope that some of you will work on them (and if so, please let me know!).

More importantly, I hope that all of you will feel sufficiently emboldened to embark on a voyage of your own choosing. All success in that endeavor.

## 8.5. Final Advice from Pasteur

> I would love to tell you that this is the best moment of my life. But I feel this happy when I finish a song or when I crack the code to a bridge that I love. For me the reward is the work – Taylor Swift on the receipt of her record-breaking fourth Grammy Award for Album of the Year (4 Feb 2024).

In his biography of Pasteur, Debré (1994) writes that in 1894, the 72 year old Pasteur met 28 year old Charles Nicolle in the hallway, asked him about himself and his work, and then advised "We must work". Debré (1994) noted Nicolle followed the advice, worked, and received the Nobel Prize in Physiology or Medicine in 1928 for his research on typhus.

We must work and, like Taylor Swift, know that the reward is the work. Bon voyage – there is much to be done and little time to be lost.

## 8.6. Summary of Major Insights

- Fruitful directions for future research include explicit models of the cyber system of the enabled physical system, incorporating human factors into the PAM and the FMSCO, and moving beyond dyadic interactions to consider multilateral cyber security agreements.

- Modeling the cyber system of the enabled physical system increases both the fidelity to operational situations and the complexity of the mathematical model but it will likely lead to new insights. Focus on a particular system, such as an electric grid and utility company is an appropriate starting point.

- Human factors are another natural extension of the PAM and FMSCO because humans are deeply involved in both creating compromise in cyber systems (e.g. by sharing thumb drives that carry malware) and detecting compromise (by recognizing anomalous situations). Modeling human factors will also expand the number of equations in the PAM and FMSCO.

- A MCSA will help movement towards a global approach to shared cyber defense and response when cyber attacks occur. In this case a specific example of a performance function related to cyber systems is national GDP as a function of national BBP, illustrated using nations that are member of the OECD. From it we learn:

  - When the likelihood of cyber attack increases with the BBP of a nation and cyber resources can be allocated from BBP to unilateral defense, thereby reducing the number of successful attacks, there is a threshold level of BBP below which no defense is predicted (because adversaries will direct attacks to nations with larger BBP). After that, resources allocated to defense increase but at a decreasing rate, so that the fraction of BBP dedicated to defense is predicted to rise, level off, and fall for the largest values of BBP.

  - Multilateral defense requires an assumption about to dependability of cyber security cooperation. When cooperation is guaranteed, nations are able to increase GDP relative to unilateral defense for all levels of BBP, with the largest relative gains going to nations with smaller BBP.

  - When cooperation is not guaranteed, we must understand a nation's perception of how likely other nations are to deliver on their commitment to the cooperative security agreement.

As a nation's perception of the trustworthiness of other nations declines, the gain from participating in a MCSA declines. Simulation methods allow us to assess the gain to a nation by participating in a MCSA and predict the fraction of nations joining the MSCA, illustrated with the OECD nations.

# Glossary of Terminology, Symbols, Equations

## Glossary of Terminology

The **attack rate parameter** $a$ multiplies the relative intensity of attack in the **Pulse Attack Model**.

The **beta probability density** is used to model a random variable whose values fall between 0 and 1. It depends upon two parameters $\alpha$ and $\beta$ and has the form $f(x|\alpha, \beta) = \frac{\Gamma(\alpha+\beta)}{\Gamma(\alpha)\Gamma(\beta)} x^{\alpha-1}(1-x)^{\beta-1}$ where $\Gamma(\cdot)$ is the **gamma function**. We can think of $\frac{\Gamma(\alpha+\beta)}{\Gamma(\alpha)\Gamma(\beta)}$ as a normalization constant.

**Broad Band Penetration (BBP)** is the number of subscriptions to fixed and mobile broadband services divided by the number of residents in a country (OECD 2017).

Cyber **co-compromise** occurs when an uncompromised cyber asset has its behavior changed significantly after interacting with a previously compromised asset.

The **co-compromise rate parameter** determines the rate at which already compromised cyber assets lead to the compromise of currently uncompromised assets.

A **cyber asset** is any kind of electronic information technology that may be operationally important in its own right or operationally important because of a linkage to other cyber assets or an enabled physical system. Cyber assets are characterized uncompromised, **compromised**, or being restored/reset from compromise.

The **co-compromise rate parameter** $a_{co}$ determines how uncompromised cyber assets are compromised by already compromised cyber assets. Due to co-compromise, cyber assets move from the uncompromised to compromised pool at rate $a_{co}x(t)x_0(t)$.

A cyber asset is **compromised** when its expected and actual behavior differ in significant ways.

A **cyber attack** occurs when compromise of a cyber asset is caused by an adversary. Cyber assets can be protected or hardened against attack, but few remain permanently invulnerable. When a cyber asset is compromised it may be repaired by humans or by other cyber assets.

To account for cyber assets that are **critical to performance**, in the simplest case, we imagine two kinds of cyber assets, now denoted by $x_1(t)$ and $x_2(t)$, with contributions to the **performance** of the cyber system $v_1$ and $v_2 > v_1$, respectively, so that the performance function when there are $x_i$ uncompromised cyber assets of each type is $\phi(x_1, x_2) = \dfrac{1}{1+e^{\frac{x_{50}-(v_1 x_1 + v_2 x_2)}{\sigma_x}}}$

When cyber assets are linked they form a **cyber system** or **cyber infrastructure**. We will not be more specific than this, keeping with the idea that cyber systems are often more easily recognized than defined.

A **Cyber Protection Team (CPT)** is a trained group of experts who both maintain defense against attack and return compromised cyber assets to their functional state.

A **cyber system** is a collection of cyber assets that act to perform a particular mission.

**Decoy** cyber assets are components of a cyber system with no functionality, but instrumented to detect compromise with high probability.

An **enabled physical system** is an electrical (e.g. communications) or mechanical (e.g. a power generator) whose performance depends upon the functioning of a focal cyber system.

**Escalation to a kinetic attack or cyber attack on critical civilian infrastructure** occurs when a nation's response to a cyber attack is a kinetic attack or cyber attack on critical civilian infrastructure.

The **exponential distribution** characterizes the positive values that a random variable $\tilde{Z}$ can take. It depends upon on parameter $\lambda$; $\tilde{Z}$ follows an exponential distribution when below the equation in the exponential distribution:

$$\Pr[\tilde{Z} \leq z] = 1 - e^{-\lambda z}$$

In its basic form the **Fundamental Model of Simultaneous Cyber Operations (FMSCO)** two adversaries, denoted by the X-side and the Y-side, conduct persistent persistent attacks at constant rates. Instead of a single network as in the **Pulse Attack Model (PAM)**, there are two interacting networks, which have similar dynamics for co-compromise and recovery as the PAM. With the assumption that the total number of cyber assets $X_T$ and $Y_T$ remain constant, there are four dynamical variables: the numbers $x(t), y(t)$ of uncompromised cyber assets of each side and the number $x_0(t), y_0(t)$ of compromised cyber assets of each side, following the dynamics

$$\frac{dx}{dt} = -axy + b(X_T - x - x_0)$$

$$\frac{dx_0}{dt} = axy - r_x x_0$$

$$\frac{dy}{dt} = -cxy + d(Y_T - y - y_0)$$

$$\frac{dy_0}{dt} = cxy - r_y y_0$$

The keys to understanding the readiness of the two cyber systems are the **rates of detection of compromise** $r_x$ (the X-side) and $r_y$ (the Y-side) at which compromised X-side and Y-side cyber assets are moved to resetting/recovery, and the **rates of recovery** $b$ (X-side) and $d$ (Y-side) at which cyber assets are moved from the resetting/recovery pool to the uncompromised pool.

The **gamma density** characterizes the values that a random variable $\tilde{V} > 0$ takes by a **shape** and **scale** parameters $a_v$ and $s_v$, respectively, with probability density $f(v) = \frac{1}{s_v^a \Gamma(a_v)} v^{a_v - 1} e^{-v/s_v}$, where $\Gamma(a_v)$ is the **gamma function**. Then mean and variance of $\tilde{V}$ are $\mathcal{E}(\tilde{V}) = a_v s_v$ and $Var(\tilde{V}) = a_v s_v^2$.

The **gamma function** is arises in classical applied mathematics and is defined by $\Gamma(a_v) = \frac{1}{s_v^a} \int_0^\infty v^{a_v - 1} e^{-v/s_v} dv$. It satisfies $\Gamma(z + 1) = z\Gamma(z)$.

**Gross Domestic Product (GDP)** is the monetary value of goods and services bought by the final user produced in a country in a given period of time.

A cyber asset is **hardened** against cyber attack when it has reduced vulnerability to the cyber attack. In general, not all cyber assets are hardened after being restored, restoring a hardened cyber asset may take longer than restoring a a more vulnerable cyber asset, and hardening is lost over time.

**Heuristic** models are not specific to any particular cyber system but have much in common with many cyber systems and allow us to increase understanding of important variables (rather to make precise predictions) developing active intuition, where "active" emphasizes that our intuition develops and grows. Heuristic models help identify what to measure to be able to assess vulnerability to cyber attack, the consequences of attack on performance of the cyber or enabled physical system, and to identify design tradeoffs and routes to defense.

The **hockey stick model** for response to cyber attack characterizes the probability that a defender responds to a cyber attack either kinetically or by an attack of critical civilian infrastructure. It is a function $p_r(\phi)$ giving the probability that the defender initiates a response given that one has not yet started when performance is $\phi$, with three parameters. There is no response if performance is greater than $\phi_1$, and maximum probability of response, $p_{max}$, when performance is less than $\phi_2$. A straight line connects those points (Figure 5.3) and the equation for the probability of response is

$$p_r(\phi) = \begin{cases} p_{max} & \phi \le \phi_2 \\ \dfrac{p_{max}}{\phi_1 - \phi_2}(\phi_1 - \phi) & \text{if } \phi_2 < \phi \le \phi_1 \\ 0 & \text{if } \phi > \phi_1 \end{cases} \tag{1}$$

A **Multilateral Cyber Security Agreement (MCSA)** is an agreement between nations to allow them to share in collective monitoring and defense against cyber attack.

The **National Trust Factor (NTF)** is the probability characterizing a nation's belief that other members of a **Multilateral Cyber Security Agreement** will honor their commitments to monitoring and defense.

The **Organization for Economic Cooperation and Development (OECD)** is an intergovernmental organization founded in 1961 to advise governments on how to develop policies that will make the lives of their citizens better. Member nations work to policy that drives reform around the world. For more information see https://www.oecd.org/en/about.html.

Research in **Pasteur's Quadrant** is motivated by an important applied problem and seeks fundamental understanding of the system of interest. It can be called applied basic research and compared to Bohr's Quadrant of pure basic research and Edison's quadrant of pure applied research.

The **performance function** of the cyber system or the enabled physical system is a value between 0 and 1, determined by the number of uncompromised cyber assets (see Mangel and McEver 2021 for an alternative), denoted by $x$, and performance is denoted by $\phi(x)$.

We assume a sigmoid function for performance $\phi(x) = \left[\dfrac{1}{1+e^{\frac{x_{50}-x}{\sigma_x}}}\right]$, where $x_{50}$ is the number of uncompromised cyber assets giving performance of 50%, and $\sigma_x$ determines how rapidly performance rises as the number of uncompromised cyber assets increases. When the performance of an enabled physical system depends upon the performance of the cyber system, we use $\mathcal{P}(\phi(x))$ to denote the performance of the physical system and assume that it too is a sigmoid.

In **process-based modeling** one describes the dynamics of the system of interest with explicit state (and often time) dependent functions of how the system changes. Such models allow us to reach beyond the empirical data, but be guided by it, and use deductive predictions to help guide data collection in the future (since the number of choices for what kind of data to collect is essentially unbounded).

A **pulse attack** characterizes the time dependence of an adversary's attack on a cyber system and is modeled using a Gaussian distribution (a bell shaped curve) with mean $t_{\text{peak}}$ (which is the time at which the pulse peaks) and dispersion (standard deviation) $\sigma$ so that the relative intensity of a pulse attack as a function of time, $I(t)$, is $I(t) = \dfrac{1}{\sqrt{2\pi}\sigma}e^{-(t_{\text{peak}}-t)^2/2\sigma^2}$. To determine the absolute intensity of the attack, we multiply relative intensity by the **attack rate parameter** $a$, so that the rate of attack at time $t$ is $aI(t)$.

In its basic form the **Pulse Attack Model (PAM)** tracks the dynamics in response to a single **pulse attack** of uncompromised $x(t)$ and compromised $x_0(t)$ assets at time $t$ in a cyber system in which the total number of cyber assets $X_T$ is constant. The dynamical equations are

$$\frac{dx}{dt} = -axI(t) - a_{co}xx_0 + b(X_T - x - x_0)$$

$$\frac{dx_0}{dt} = axI(t) + a_{co}xx_0 - r_xx_0$$

where $a$ is the **attack rate parameter**, $I(t)$ is the **relative intensity of a pulse attack**, and $a_{co}$ is the **rate of co-compromise**.

The **relative intensity of a pulse attack** is modeled by as Gaussian distribution $I(t) = \frac{1}{\sqrt{2\pi}\sigma} e^{-(t_{\text{peak}}-t)^2/2\sigma^2}$.

The **resilience stack** captures the hierarchical nature of attack and defense in cyber systems, in the sense that the best defense is to avoid attack in the first place, but when attack cannot be avoided it can be resisted; when resistance is unsuccessful and anticipating that the attacker enters the cyber system, the cyber system can be hardened so that it is robust to attack or may have redundant cyber assets, so that if one cyber asset fails others can take over the mission. Since even hardened, redundant systems may fail against some sophisticated cyber tools assets will need to be repaired, reset, or even replaced.

A compromised cyber asset is **restored** or **reset** after compromise is detected (usually requiring some time) and the cyber asset is repaired sufficiently to return to operational status.

**Resources** are rate parameters such as $b$ and $r_x$ in the **PAM** or $b, r_x, d$ and $r_y$ in the **FMSCO** ostensibly under the control of the defender, who can choose their values, subject to a constraint such as $c_b b + c_r r_x = \mathcal{R}$ for the PAM, where $c_b$ and $c_r$ are costs of a unit of $b$ and $r_x$, respectively, and $\mathcal{R}$ is the total resource level.

**Simultaneous Cyber Operations (SCOs)** occur when multiple actors conduct cyber operations against each other, resulting in continuous action in cyberspace.

The **SIRS** and **SIR** models are classical ones in the population biology of disease, characterizing the dynamics of Susceptible ($S$), Infected ($I$), and recovered individuals. In the SIRS model, recovered individuals lose their immunity to the disease as time progresses and once more become susceptible to it. When there is no loss of immunity we have the SIR model. Letting $t$ denote time, the symbols $S(t), I(t)$, and $R(t)$ denote the number of Susceptible, Infected, and

Recovered individuals at time $t$. If total population size is constant, they follow the dynamics

$$\frac{dS}{dt} = -\beta I(t)S(t) + \gamma R(t)$$

$$\frac{dI}{dt} = \beta I(t)S(t) + \mu I(t)$$

$$\frac{dR}{dt} = \mu I(t) - \gamma R(t)$$

In **stochastic versions** of the PAM and the FMSCO, we introduce random components to the dynamics, so that instead of steady states and single numbers for quantities such as numbers of uncompromised cyber assets, performance, and recovery time, we obtain distributions.

The **Stochastic Simulation Algorithm** (SSA) and $\tau$-**Leaping Algorithm** are two ways of developing stochastic versions of the **Pulse Attack Model** and **Fundamental Model of Simultaneous Cyber Operations** in which the terms on the right hand side of ordinary differential equation are used to compute the probability of change in the relevant state variable.

# Bibliography

Abramowitz, M. and I.A. Stegun. 1965. *Handbook of Mathematical Functions*. Dover Publications, New York.

Agutter, P.S. and J.A. Tuszynski. 2011. Analytic theories of allometric scaling. *The Journal of Experimental Biology* 214:1055–1062.

Alpcan, T. and T. Basar. 2011. *Network Security. A Decision and Game-Theoretical Approach*. Cambridge University Press, Cambridge, UK.

Ambrus, M., Arts, K., Hey, E., and H. Raulus. 2014. *The Role of 'Experts' in International and European Decision-Making Processes. Advisors, Decision Makers or Irrelevant Actors?*. Cambridge University Press, Cambridge, UK.

Anderson, R.M. and R.M. May. 1991. *Infectious Diseases of Humans. Dynamics and Control*. Oxford University Press, Oxford, UK.

Andresen, S. 2014. The role of scientific expertise in multilateral environmental agreements: Influence and effectiveness. In *The Role of 'Experts' in International and European Decision-Making Processes. Advisors, Decision Makers or Irrelevant Actors?* (pp. 105–125). Ambrus, M., Arts, K., Hey, E., and H. Raulus (editors). Cambridge University Press, Cambridge, UK.

Arquilla, J. and D. Ronfeldt (editors). 1997. *In Athena's Camp. Preparing for Conflict in the Information Age*. RAND Corporation, Monograph Reports (MR) 880, Santa Monica, CA.

Aurell, A., Carmona, R., Dayankli, G., and M. Lauriere. 2022. Optimial incentives to mitigate epidemics: A Stackelberg mean field game approach. *SIAM Journal on Optimal Control* 60:S294–S322.

Axelrod, R. 1984/2006 (revised edition). *The Evolution of Cooperation*. Basic Books, New York.

Axelrod, R. 2014. A repertory of cyber analogies. In *Cyber Analogies*, (pp. 108–117). Goldman, E.O. and Arquilla, J. (editors). Technical Report NPS-DA-14-001, Naval Postgraduate School, Monterey, CA.

Axelrod, R. and R. Iliev. 2014. Timing of cyber conflict. *Proceedings of the National Academy of Sciences US* 111:1298–1303.

Aylward, B. and R. Tangermann. 2011. The global polio eradication initiative: Lessons learned and prospects for success. *Vaccine* 295: D80–D85.

Bangham, C.R.M. and B. Asquith. 2001. Theoretical Biology: Viral immunology from math. *Science* 291:992–994.

Barnett, V.D. 1962. The Monte Carlo solution of a competing species problem. *Biometrics* 18:76–103.

Barrett, S. 2003. *Environment & Statecraft*. Oxford University Press, Oxford, UK.

Bartlett, M.S., Gower, J.C., and P.H. Leslie. 1960. A comparison of theoretical and empirical results for some stochastic population models. *Biometrika* 47:1–11.

Basar, T. and G.J. Olsder. 1982. *Dynamic Noncooperative Game Theory*. Academic Press, London.

Bazykin, A.D. 1998. *Nonlinear Dynamics of Interacting Populations*. World Scientific, Singapore.

Bellovin, S., Landau, S., and H. Lin. 2018. Limiting the undesired impact of cyber weapons: Technical requirements and policy implications. In *Bytes, Bombs, and Spies*. (pp. 265–288). Lin, H. and A. Zegart (editors). Brookings Institution Press, Washington, DC.

Benvenisti, E. and G. Nolte. 2018. *Community Interests Across International Law*. Oxford University Press, Oxford, UK.

Berry, R.J. 1989. Ecology: Where genes and geography meet. *Journal of Animal Ecology* 58:733–759.

Bertsekas, D.P. *Dynamic Programming and Optimal Control. Volume II*. Athena Scientific, Belmont, MA, USA.

Bhunia, B., Kar, T.K., and P. Debnath. 2023. Explicit impacts of harvesting on a delayed predator-prey system with Allee effect. *International Journal of Dynamics and Control*. https://link.springer.com/article/10.1007/s40435-023-01167-9. Last accessed 10 April 2025.

Binmore, K. 1994. *Game Theory and the Social Contract Volume I*. The MIT Press, Cambridge, MA.

Bjørnstadt, P., Shea, K., Krzywinski, M. and N. Altman. 2020a. Modeling infectious epidemics. *Nature Methods* 17:455–456.

Bjørnstadt, P., Shea, K., Krzywinski, M. and N. Altman. 2020b. The SEIRS model for infectious disease dynamics. *Nature Methods* 17:555–558.

Borghard, E.D. and S.W. Lonergan. 2017. The logic of coercion in cyberspace. *Security Studies* 26:452–481.

Borghard, E.D. and S.W. Lonergan. 2019. Cyber operations as imperfect tools of escalation. *Strategic Studies Quarterly* 13:122–145.

Brouwer, A.F., Eisenberg, M/C., Love, N.G., and N.S. Eisenberg. 2019. Phenotypic variations in persistence and infectivity between and within environmentally transmitted pathogen populations impact population-level epidemic dynamics. *BMC Infectious Diseases* 19:449, doi:10.1186/s12879-019-4054-8.

Brunnèe, J. 2018. International environmental law and community interests. In *Community Interests Across International Law* (pp. 151–175). E. Benvesti and G. Nolte (editors). Oxford University Press, Oxford, UK.

Buchanan, B. 2016. *The Cybersecurity Dilemma*. Oxford University Press, Oxford, UK.

Budiansky, S. 2013. *Blackett's War*. A.A. Knopf, New York.

Burris, C.M., McEver, J.G., Schoenborn, H.W., and D.T. Signori. 2010. Steps toward improved analysis for network mission assurance. *IEEE International Conference on Social Computing/IEEE International Conference on Privacy, Security, Risk and Trust*, pp. 1177–1182, doi:10.1109/SocialCom.2010.175.

Carlin J.P.(with G.M. Graff). 2018. *Dawn of the Code War and the Rising Global Cyber Threat*. Hachette Book Group, New York.

Caswell, H. 1988. Theory and models in ecology: A different perspective. *Ecological Modelling* 43:33–44.

Chamberlain, T.C. 1897. Studies for students: The method of multiple working hypotheses. *The Journal of Geology* 5:837–848.

Chayes, A. and A.H. Chayes. 1995. *The New Sovereignty. Compliance with International Regulatory Agreements*. Harvard University Press, Cambridge, MA.

Chumakov, K., Ehrenfeld, E., Agol, V.I., and E. Wimmer. 2021. Polio eradication at the crossroads. *Lancet Global Health* 9:e1172–1175.

Clark, C.W. and M. Mangel. 2000. *Dynamic State Variable Models in Ecology. Methods and Applications*. Oxford University Press, Oxford and New York.

Cooke, K.L. and Z. Grossman. 1982. Discrete delay, distributed delay and stability switches. *Journal of Mathematical Analysis and Its Applications* 86:592–627.

Coveney, P. and R. Highfield. 2023. *Virtual You. How Building Your Digital Twin Will Revolutionize Medicine and Change Your Life*. Princeton University Press, Princeton, NJ.

Crease, R.P. 2008. *The Great Equations. Breakthroughs in Science from Pythagoras to Heisenberg*. W.W. Norton and Company, New York.

Danzig, R.J. 2014. Surviving on a diet of poisoned fruit. Reducing the national security risks of America's cyber dependencies. Center for a New American Security. Washington, DC. Available at https://www.

cnas.org/publications/reports/surviving-on-a-diet-of-poisoned-fruit-reducing-the-national-security-risks-of-americas-cyber-dependencies. Last accessed 10 April 2025.

Davis, H.T. 1962. *Introduction to Nonlinear Differential and Integral Equations*. Dover Publications, New York.

Davis, P.K. 2000. Exploratory analysis enabled by multiresolution, multiperspective modeling. In *2000 Winter Simulation Conference Proceedings (Cat. No.00CH37165), Orlando, FL, USA, 2000*, pp. 293–302, vol.1, doi:10.1109/WSC.2000.899731. This document is also published, with authors P.K. Davis, J.H. Bigelow, and J. McEver as Rand Research Report RP-925. Freely available at https://www.rand.org/pubs/reprints/RP925.html. Last accessed 10 April 2025.

Debré, P. 1994. *Louis Pasteur*. Johns Hopkins University Press, Baltimore.

Dell Technologies 2023. *How Higher Ed is Answering the Cybersecurity Challenge*. Available at https://www.delltechnologies.com/asset/en-us/solutions/industry-solutions/industry-market/how-higher-ed-answers-cybersecurity-challenge.pdf. Last accessed 10 April 2025.

Dipert, R.R. 2010. The ethics of cyberwarfare. *Journal of Military Ethics*, 9(4):384–410, doi:10.1080/15027570.2010.536404.

Dixit, A.K. and R.S. Pindyck. *Investment under Uncertainty*. Princeton University Press, Princeton, NJ.

Driver, R.D. 1977. *Ordinary and Delay Differential Equations*. Springer Verlag, New York.

Dwyer, G., Dushoff, J., Elkinton, J.S., and S.A. Levin. 2000. Pathogen-driven outbreaks in forest defoliators revisited: Building models from experimental data. *The American Naturalist* 156:105–120.

Dye, C. 1984. Models for the population dynamics of the Yellow Fever mosquito, *Aedes aegypti*. *Journal of Animal Ecology* 53:247–268.

Easley, D. and J. Kleinberg. 2010. *Networks, Crowds, and Markets: Reasoning about a Highly Connected World*. Cambridge University Press, Cambridge, UK.

Edelstein-Keshet, L. 1988. *Mathematical Models in Biology*. Random House, New York.

Epstein, J. 1997. *Nonlinear Dynamics, Mathematical Biology, and Social Science*. Addison-Wesley Publishing Company, Reading, MA.

Farmelo, G. (editor) 2002. *It Must be Beautiful. Great Equations of Modern Science*. Granta Books, London and New York.

Farrell, H. and C.L. Glaser. 2018. How effects, saliences, and norms should influence U.S. cyberwar doctrine. In *Bytes, Bombs, and Spies* (pp. 45–79). Lin H. and A. Zegart (editors). Brookings Institution Press, Washington, DC.

Farsangi, E.N., Takewaki, I., Yang, T.Y., Astaneh-Asl, A., and P. Gardoni (editors) 2019. *Resilient Structures and Infrastructure*. Springer Nature Singapore Pte., Singapore.

Feller, W. 1968. *An Introduction to Probability Theory and Its Applications, Volume 1*. Wiley Interscience, New York.

Fenner, F. 1982. Global eradication of smallpox [with Discussion]. *Reviews of Infectious Diseases* 4:916–930.

Feynman, R.P. 1985. *"Surely You're Joking Mr. Feynman!" Adventures of a Curious Character*. W.W. Norton, New York.

Filatrella, G., Nielsen, A.H., and N.F. Pedersen. 2008. Analysis of a power grid using a Kuramoto-like model. *European Physical Journal B* 61: 485–491.

Fischerkeller, P.M., Harknett, R.J. and J. Vićić, 2020. The limits of deterrence and the need for persistence. In *The Cyber Deterrence Problem* (pp. 21–38). Brantly, A.F. (editor). Rowman & Littlefield International, London, UK

Gartzke, E. 2013. The myth of cyberwar: Bringing war in cyberspace back down to earth. *International Security* 38(2):41–73.

Gartzke, E. and J.R. Lindsay. 2015. Weaving tangled webs: Offense, defense, and deception in cyberspace. *Security Studies* 24:316–348.

Gause, G.F. 1934/2019. *The Struggle for Existence*. Dover Publications (2019 edition), Mineola, NY.

Gillespie, D.T. 1977. Exact stochastic simulation of coupled chemical reactions. *Journal of Physical Chemistry* 81:2340–2361.

Gillespie, D.T. 2001. Approximate accelerated stochastic simulation of chemically reacting systems. *Journal of Chemical Physics* 115:1716–1733.

Gillespie, D.T. 2007. Stochastic simulation of chemical kinetics. *Annual Review Physical Chemistry* 58:35–55.

Gillespie, D.T. and M. Mangel. 1981. Conditioned averages in chemical kinetics. *Journal of Chemical Physics* 75:704–709.

Goldman, E.O. 2020. The cyber paradigm shift. In *Ten Years In: Implementing Strategic Approaches to Cyberspace* (pp. 31–45). Schneider, J.G., Goldman, E.O., and M. Warner, (editors). US Naval War College Newport Papers 45.

Gomez, M.A. and C. Whyte. 2021. Breaking the myth of cyber doom: Securitization and normalization of novel threats. *International Studies Quarterly* 65:1137–1150, doi:10.1093/isq/sqab034.

Goodstein, J.R. 2007. *The Volterra Chronicles. The Life and Times of an Extraordinary Mathematician 1860–1940*. American Mathematical Society, Providence Rhode Island.

Green, B.R. and A. Long. 2019. Conceal or reveal? Managing clandestine military capabilities in peacetime competition. *International Security* 44:48–83.

Gross, M.L., Canetti, D., and D.R. Vashdi. 2018. Cyber terrorism. Its effects on psychological well-being, public confidence, and political attitudes.

In *Bytes, Bombs, and Spies* (pp. 235–265). Lin, H. and A. Zegart (editors). Brookings Institution Press, Washington, DC.

Gupta, J. 2014. Global scientific assessments and environmental resource governance: Towards a science-policy interface ladder. In *The Role of 'Experts' in International and European Decision-Making Processes. Advisors, Decision Makers or Irrelevant Actors?* (pp. 148–170). Ambrus, M., Arts, K., Hey, E., and H. Raulus (editors). Cambridge University Press, Cambridge, UK

Guzman, A.T. 2008. *How International Law Works. A Rational Choice Theory.* Oxford University Press, Oxford, UK.

Hackett, S.C. and M.B. Bonsall. 2018. Management of a stage-structured insect pest: An application of approximate optimization. *Ecological Applications* 28:938–95.

Hackett, S.C. and M.B. Bonsall. 2019. Insect pest control, approximate dynamic programming, and the management of the evolution of resistance. *Ecological Applications*, 29:e01851.

Hartman, P. 1973. *Ordinary Differential Equations.* Johns Wiley and Sons, Baltimore.

Healey, J. 2018. The Cartwright conjecture. In *Bytes, Bombs, and Spies* (pp. 173–194). H. Lin and A. Zegart (editors). Brookings Institution Press, Washington, DC.

Hechter, M. 1987. *Principles of Group Solidarity.* University of California Press, Berkeley.

Henderson, D.A. 1987. Principles and lessons from the smallpox eradication programme. *Bulletin of the World Health Organization* 65:535–546.

Henderson, D.W. and D. Taimina. 2020. *Experiencing Geometry. Euclidean and Non-Euclidean Geometry with History, fourth edition.* Project Euclid. Open access at https://projecteuclid.org. Last accessed 10 April 2025.

Herzog, M. 2003. Introduction. In *The War of Atonement. The Inside Story of the Yom Kippur War* (pp. xi–xvii). Herzog, C. (editors). Greenhill Books, London.

Hilborn, R. and M. Mangel. 1997. *The Ecological Detective. Confronting Models with Data.* Princeton University Press, Princeton, NJ.

Hobbs, M.T. and M.B. Hooten. 2015. *Bayesian Models. A Statistical Primer for Ecologists.* Princeton University Press, Princeton, N.J.

Huth, P. and B. Russett. 1988. Deterrence failure and crisis escalation. *International Studies Quarterly* 32:29–45.

Ingelsby, T.V., O'Toole, T., Henderson, D.A., *et al.* 2002. Anthrax as a biological weapon, 2002. Updated recommendations for management. *Journal of the American Medical Association* 288:2236–2252.

Janis, I.L. and L. Mann. 1976. Coping with decisional conflict. *American Scientist* 64:657–667.

Janis, I.L. and L. Mann. 1977. *Decision Making*. The Free Press, Macmillan Publishing Co, New York.

Johnson, A.W., Duda, K.R., Sheridan, T.B., and C.M. Oman. 2017. A closed-loop model of operator visual attention, situation awareness, and performance across automation mode transitions. *Human Factors* 59:229–241.

Kadowaki, K. 2023. A primer of community ecology using the R language. *Population Ecology* 65:240–256.

Kahneman, D. 2012. *Thinking, Fast and Slow*. Farrar, Strauss, and Giroux, New York.

Kapczynski, A. 2016. Order without intellectual property law: Open science in influenza. *Cornell Law Review* 102:1539–1615.

Keane, T. 2011. Combat modelling with partial differential equations. *Applied Mathematical Modelling* 35:2723–2735.

Keeling, M.J. and P. Rohani. 2008. *Modeling Infectious Diseases in Humans and Animals*. Princeton University Press, Princeton, NJ.

King, A. and M. Gallagher (co-chairs) 2020. *Report of the U.S. Cyberspace Solarium Commission*. Available at https://www.solarium.gov/report. Last accessed 10 April 2025.

Koch, K-R. 2007. *Introduction to Bayesian Statistics*, 2nd edition. Springer Verlag, Berlin Heidelberg.

Koppel, T. 2015. *Lights Out. A Cyberattack. A Nation Unprepared. Surviving the Aftermath*. Random House, New York.

Kostitzin, V.A. 1939. *Mathematical Biology*. George G. Harrap & Company, London.

Kostyuk, N. 2021. Deterrence in the cyber realm: Public versus private cyber capacity. *International Studies Quarterly* 65:1151–1162, doi:10.1093/isq/sqab039.

Kostyuk, N., Powell, S., and M. Skach. 2018. Determinants of the cyber escalation ladder. *The Cyber Defense Review* 3(1):123–134.

Krener, A.J. 1979. A formal approach to stochastic integration and differential equations. *Stochastics* 3:105–125.

Kreps, S. and J. Schneider. 2019. Escalation firebreaks in the cyber, conventional, and nuclear domains: Moving beyond effects-based logics. *Journal of Cybersecurity* 5(1):tyz007, doi:10.1093/cybsec/tyz007.

Lachow, I. and T. Grossman. 2018. Cyberwar Inc. Examining the role of companies in offensive cyber operations. In *Bytes, Bombs, and Spies* (pp. 379–399). Lin, H. and A. Zegart (editors). Brookings Institution Press, Washington, DC

Lafferty, K.D., Smith, K.F., and E.M.P. Madin. 2008. The infectiousness of terrorist ideology. In *Natural Security. A Darwinian Approach to a Dangerous World* (pp. 186–206). Sagarin R.D. and T. Taylor (editors). University of California Press, Berkeley and Los Angeles.

Lanchester, F.W. 1917. *Aircraft in Warfare. The Dawn of the Fourth Arm.* D. Appleton & Company, New York.

Lebow, R.N. and J.G. Stein. 1990. Deterrence: The elusive dependent variable. *World politics*, 42:336–369.

Leslie, P.H. and J.C. Gower. 1958. The properties of a stochastic model for two competing species. *Biometrika* 45:316–330.

Levin, S. 1998. Ecosystems and the biosphere as complex adaptive systems. *Ecosystems* 1:431–436.

Levin S. 1999. *Fragile Dominion. Complexity and the Commons.* Helix Books, Reading, MA.

Levin, S. 2003. Complex adaptive systems: Exploring the known, the unknown, and the unknowable. *Bulletin of the American Mathematical Society* 40:3–19.

Libicki, M.C. 2016. *Cyberspace in Peace and War.* Naval Institute Press, Annapolis, MD.

Libicki, M.C. 2018. Second acts in cyberspace. In *Bytes, Bombs, and Spies* (pp. 133–149). Lin, H. and A. Zegart (editors). Brookings Institution Press, Washington, DC.

Libicki, M.C., 2018a. Expectations of cyber deterrence. *Strategic Studies Quarterly* 12:44–57.

Libicki, M.C., Ablon, L., and T. Webb. 2015. *The Defender's Dilemma. Charting a Course Toward Cybersecurity.* Rand Corporation, Santa Monica, CA.

Lin, H. and A. Zegart (editors) 2019. *Bytes, Bombs, and Spies.* Brookings Institution Press, Washington, DC.

Lindsay, J.R. and E. Gartzke. 2018. Coercion through Cyberspace. The stability-instability paradox revisited. In *Coercion. The Power to Hurt in International Politics* (pp. 179–203). K.M. Greenhill and P. Krause (editors). Oxford University Press, New York.

Link, J., Huse, G., Gaichas, S. and A.R. Marshak. 2030. Changing how we approach fisheries: A first attempt at an operational framework for ecosystem approaches to fisheries management. *Fish and Fisheries* 21:393–434.

Liu, S., Mashayekh, S., Kundar, D., Zourntos, T., and K. Butler-Purry. 2013. A framework for modeling cyber-physical switching attacks in smart grid. *IEEE Transactions on Emerging Topics in Computing* 1:273–285.

Long. A. 2018. A cyber SIOP? Operational considerations for strategic offensive cyber planning. In *Bytes, Bombs, and Spies* (pp. 105–132). Lin, H. and A. Zegart (editors). Brookings Institution Press, Washington, DC.

Lotka, A.J. 1924/1956. *Elements of Mathematical Biology.* Dover Publications (1956 edition), New York.

MacDonald, D. 1989. *Biological Delay Systems: Linear Stability Theory.* Cambridge University Press, Cambridge, UK.

Maness, R.C. and B. Valeriano. 2016. The impact of cyber conflict on international interactions. *Armed Forces & Society* 42:301–323.

Mangel, M. 1979. Fluctuations in systems with multiple steady states. Applications to Lanchester equations. In *Information Linkage Between Applied Mathematics and Industry* (pp. 335–346). P. Wang (editor). Academic Press, New York.

Mangel, M. 1982. Applied mathematicians and naval operators. *SIAM Review* 24:289–300.

Mangel, M. 1985. *Decision and Control in Uncertain Resource Systems.* Academic Press, San Diego and New York.

Mangel, M. 1994. Barrier transitions driven by fluctuations, with applications to ecology and evolution. *Theoretical Population Biology* 45:16–40.

Mangel, M. 2006. *The Theoretical Biologist's Toolbox.* Cambridge University Press, Cambridge, UK.

Mangel, M. 2015. Stochastic dynamic programming illuminates the link between environment, physiology, and evolution. *Bulletin of Mathematical Biology* 77:857–877.

Mangel, M. 2017. Know your organism, Know your data. *ICES Journal of Marine Science*, doi:10.1093/icesjms/fsw228.

Mangel, M. 2023. Pest science in pasteurs quadrant. *Journal of Pest Science*, doi:10.1007/s10340-023-01633-5.

Mangel, M. and D. Ludwig. 1977. Probability of extinction in a stochastic competition. *Society for Industrial and Applied Mathematics (SIAM) Journal on Applied Mathematics* 33:256–266.

Mangel, M. and M.B. Bonsall. 2008. Phenotypic evolutionary models in stem cell biology: replacement, quiescence, and variability. *PLoS ONE* 3(2):e1591, doi:10.1371/journal.pone.0001591.

Mangel, M. and J. McEver. 2021. Modeling coupled nonlinear multilayered dynamics: Cyber attack and disruption of an electric grid. *Complexity*, doi:10.1155/2021/5584123.

Mangel, M. and A. Brown. 2022. Population processes in cyber system variability. *PLoS ONE* 17(12):e0279100, doi:10.1371/journal/pone.0279100.

Mangel, M. *et al.* 1996. Principles for the conservation of wild living resources. *Ecological Applications* 6:338–362.

May, R.M. 2002. The best possible time to be alive. The logistic map. In *It Must be Beautiful. Great Equations of Modern Science* (pp. 212–228). Farmelo, G. (editor). Granta Books, London and New York.

Maynard Smith, J. 1982. *Evolution and the Theory of Games.* Cambridge University Press, Cambridge, UK.

Maynard Smith, J. 2002. Equations of life. The mathematics of evolution. In *It Must be Beautiful. Great Equations of Modern Science* (pp. 193–211). Farmelo, G. (editor). Granta Books, London and New York.

McCue, B. 2020. *Beyond Lanchester. Stochastic Granular Attrition Combat Processes.* Self-published; available at Amazon.com.

McCue, B. 2022. *Essays on Search Theory: With Examples From The U-Boat War.* Available from Amazon.

McDermott, R. 2010. Decision making under uncertainty. In National Research Council 2010. *Proceedings of a Workshop on Deterring Cyberattacks: Informing Strategies and Developing Options for U.S. Policy* (pp. 227–241). The National Academies Press, Washington, DC. doi: 10.17226/12997.

McDonald, N. 1989. *Biological Delay Systems: Linear Stability Theory.* Cambridge University Press, Cambridge, UK.

McElreath, R. 2020. *Statistical Rethinking: A Bayesian Course with Examples in R and STAN*, 2nd edition. CRC Press, Boca Raton, FL.

McElreath, R. and R. Boyd. 2007. *Mathematical Models of Social Evolution.* University of Chicago Press.

McEver, J. Vasatka, J., Syphert, R., West, T.D., and Z. Ben Miled. 2019. Cybersecurity as a complex adaptive systems problem. In *Complex Systems Engineering: Theory and Practice* (pp. 201–213). Flumerfelt, S. Schwartz, K.G. Marvis, D. and S. Briceno (editors). American Institute of Aeronautics and Astronautics.

McNamara, J.M. and O. Leimar. 2020. *Game Theory in Biology.* Princeton University Press, Princeton, NJ.

McPeek, M.A. 2017. *Evolutionary Community Ecology.* Princeton University Press, Princeton, NJ.

McPeek, M.A. 2022. *Coexistence in Ecology. A Mechanistic Perspective.* Princeton University Press, Princeton, NJ.

McPeek, M.A., McPeek, S.J., and J.L. Bronstein. 2022. Nectar dynamics and the coexistence of two plants that share a pollinator. *Oikos* 2022:e08869, doi:10.1111/oik.08869.

Merl, D., Johnson, L.R., Gramacy, R.B., and M. Mangel M. 2009. A statistical framework for the adaptive management of epidemiological interventions. *PLoS ONE* 4(6):e5807, doi:10.1371/journal.pone.0005807.

Merrill, T.E.S., Rapti, Z., and C.E. C/'aceres. 2021. Host controls within-host disease dynamics: Insight from an invertebrate system. *American Naturalist* 198:317–332.

Meyer, J. 2024. On the need to understand human behavior to do analytics of behavior. In *Knowledge and Digital Technology* (pp. 47–62). Glückler, J. and R. Panitz (editors). Springer Verlag, Cham Switzerland.

Microsoft. 2022. Defending Ukraine: Early Lessons from the Cyber War. 22 June 2022. Available at https://blogs.microsoft.com/on-the-issues/2022/06/22/defending-ukraine-early-lessons-from-the-cyber-war/. Last accessed 10 April 2025.

Microsoft 2023. Volt Typhoon targets US critical infrastructure with living-off-the-land techniques. Available at https://www.microsoft.com/en-us/security/blog/2023/05/24/volt-typhoon-targets-us-critical-infrastructure-with-living-off-the-land-techniques/ Last accessed 10 April 2025.

Miller, A.D., Roxburgh, S.H., and K. Shea. 2011. How frequency and intensity shape diversity-disturbance relationships. *Proceedings of the National Academy of Sciences I\*S* 108:5643–5648.

Miller, A.D., Deilly, D., Bauman, S., and K. Shea. 2012. Interactions between frequency and size of disturbance affect competitive outcomes. *Ecological Research* 27:783–791.

Miller, L. 2020. *Ransomeware Defense for Dummies. Cisco*, 2nd special edition. John Wiley & Sons, Hoboken, NJ.

Morse, P.M. 1977. *In at the Beginnings: A Physicist's Life* MIT Press, Cambridge, MA.

Murdoch, W.W., Nisbet, R.M., Blythe, S.P., Gurney, W.S.C., and J.D. Reeve. 1987. An invulnerable age class and stability in delay-differential parasitoid-host models. *The American Naturalist* 129:263–282.

Murdoch, W.W., Briggs, C.J., and R.M. Nisbet. 2003. *Consumer-Resource Dynamics*. Princeton University Press, Princeton, NJ.

Murray, J.D. 2002. *Mathematical Biology. I. An Introduction*, third edition. Springer Verlag, New York.

Nakasone, P.M. 2020. A cyber force for persistent operations. In *Ten Years In: Implementing Strategic Approaches to Cyberspace* (pp. 1–7). Schneider, J.G., Goldman, E.O., and M. Warner (editors). US Naval War College Newport Papers 45.

National Academies of Sciences, *Engineering, and Medicine 2017. Enhancing the Resilience of the Nation's Electricity System*. The National Academies Press, Washington, DC. doi:10.17226/24836

National Research Council 2010. *Proceedings of a Workshop on Deterring Cyberattacks: Informing Strategies and Developing Options for*

*U.S. Policy.* The National Academies Press, Washington, DC. doi:10.17226/12997.

National Research Council 2013. *The Resilience of the Electric Power Delivery System in Response to Terrorism and Natural Disasters: Summary of a Workshop.* The National Academies Press, Washington, DC. doi:10.17226/18535.

NESCOR. 2013. Electric Sector Failure Scenarios and Impact Analyses. Available at https://smartgrid.epri.com/doc/NESCOR%20failure%20 scenarios09-13%20finalc.pdf. Last accessed 10 April 2025.

Nye, J.S. 2010. *Cyber Power.* Report, Belfer Center for Science and International Affairs. May 2010. Available at https://www.belfercenter.org/ publication/cyber-power. Last accessed 10 April 2025.

Nye, J.S. 2011. Nuclear lessons for cyber security? *Strategic Studies Quarterly* 5(4):18–38.

Nye, J.S. 2017. Deterrence and dissuasion in cyberspace. *International Security* 41:44–71.

OECD. 2017. OECD Science, *Technology and Industry Scoreboard 2017: The Digital Transformation.* OECD Publishing, Paris, doi: 10.1787/9789264268821-en.

Parsons, T.L. 2018. Invasion probabilities, hitting times, and some fluctuation theory for the stochastic logistic process. *Journal of Mathematical Biology* 77:1193–1231.

Pearson, A. 2017. *How Hard Can It Be?* St. Martin's Press, New York.

Perkovich, G. and A.E. Levite (editors). 2017. *Understanding Cyber Conflict. 14 Analogies.* Georgetown University Press, Washington, DC.

Pilowsky, J.A., Colwell, R.K., Rahbek, C., and D.A. Fordham. 2022. Process-explicit models reveal the structure and dynamics of biodiversity patterns. *Science Advances* 8:eabj2271.

Pirolli, P. 2009. *Information Foraging Theory. Adaptive Interaction with Information.* Oxford University Press, Oxford and New York.

Powell, W.B. 2011. *Approximate Dynamics Programming. Solving the Curses of Dimensionality.* Wiley. New York.

Prisig, R.M. 1974/1999. *Zen and the Art of Motorcycle Maintenance.* William Morrow & Company, New York.

Punt, A.E., A'mar T., Bond, N.A., *et al.* 2014. Fisheries management under climate and environmental uncertainty: control rules and performance simulation. *ICES Journal of Marine Science* 71:2208–2220.

Råberg, L., Graham, A.L., and A.F., Read. 2009. Decomposing health: tolerance and resistance to parasites in animals. *Philosophical Transactions of the Royal Society B* 364:37–49.

Ramsay, K.W. 2017. Information, uncertainty, and war. *Annual Review of Political Science* 20:505–527.

Read, A.F., Graham, A.L., and L. Råberg. 2008. Animal defenses against infectious agents: is damage control more important than pathogen control? *PLOS Biology* 6:2638–2641.

Reames, R. 2024. *The Ancient Art of Thinking for Yourself. The Power of Rhetoric in Polarized Times.* Basic Books, New York.

Rid, T. 2012. Cyber war will not take place. *Journal of Strategic Studies* 35(1):5–32.

Rid, T. and B. Buchanan. 2015. Attributing syver attacks. *Journal of Strategic Studies* 38(1–2):4–37.

Roopnarine, P.D., Banker, R.M.W., and S.D. Sampson. 2022. Impact of the extinct megaherbivore Steller's sea cow (*Hydrodamalis gigas*) on kelp forest resilience. *Frontiers in Ecology and Evolution*, doi:10.3389/fevo.2022.983558.

Rovner, J. 2020. Cyberspace and warfighting. In *Ten Years In: Implementing Strategic Approaches to Cyberspace* (pp. 81–96). Schneider, J.G., Goldman, E.O., and M. Warner (editors). US Naval War College Newport Papers 45.

Ruxton, G.D., Allen, W.L., Sherratt, T.N., and M.S. Speed. 2018. *Avoiding Attack. The Evolutionary Ecology of Crypsis, Aposematism, and Mimicry*, second edition. Princeton University Press, Princeton, NJ.

Sage, A.P. Behavioral and organizational considerations in the design of information systems and processes for planning decision support. *IEEE Transactions on Systems, Man, and Cybernetics* SMC11:640–678.

Sandler, T. 2004. *Global Collective Action*. Cambridge University Press, Cambridge, UK.

Schmitt, M.N. and L. Vihul (editors) 2017. *Tallinn Manual 2.0 on the International Law Applicable to Cyber Operations*. Cambridge University Press, Cambridge, UK.

Schneider, J.G. 2020. Cyber strategy, talent, and great power competition. In *Ten Years In: Implementing Strategic Approaches to Cyberspace*, (pp. 141–157). Schneider, J.G., Goldman, E.O., and M. Warner (editors). US Naval War College Newport Papers 45.

Schneider, J.G., Goldman, E.O., and M. Warner (editors). 2020. Ten Years In: *Implementing Strategic Approaches to Cyberspace*. US Naval War College Newport Papers 45.

Segal, A. 2018. U.S. Offensive Cyber Operations in a China-U.S. Military Confrontation. In *Bytes, Bombs, and Spies* (pp. 319–342). Lin, H. and A. Zegart (editors). Brookings Institution Press, Washington, DC.

Serevido, M.R., Brandvain, Y., Dhole, S., Fitzpatrick, C.L., Goldberg, E.E., Stern, C.A., Van Cleve, J., and D.J. Yeh. 2014. Not just a theory? The utility of mathematical models in evolutionary biology. *PLoS Biol* 12(12):e1002017, doi:10.1371/journal.pbio.

Sheldon, R. 2016. Military Operations Research Society (MORS) oral history project interview of Dr. Phil E. DePoy. *Military Operations Research* 21(2):7–36, doi:10.5711/1082598321207.

Sherrat, T.N. and A. Stefan. 2024. Capture tolerance: A neglected third component of aposematism. *Evolutionary Ecology*, doi:10.1007/s10682-024-10289-1.

Shou, W., Bergstrom, C.T., Chakraborty, A.K., and F.K. Skinner. 2015. Theory, models and biology. *eLife* 4:e07158, doi:10.7554/eLife.07158.

Singer, P.W. and A. Friedman. 2014. *Cybersecurity and Cyberwar. What Everyone Needs to Know.* Oxford University Press, Oxford and New York.

Siple, M.C., Essington, T.E., and É.E. Plagányi. 2018. Forage fish fisheries management requires a tailored approach to balance trade-offs. *Fish and Fisheries* 20:110–124.

Skyrms, B. 2004. *The Stag Hunt and the Evolution of Social Structure.* Cambridge University Press, Cambridge, UK.

Slater, R. 2017. What is the cyber offense-defense balance? *International Security* 41:72–109.

Slayton, R. 2016/2017. What is the cyber offense-defense balance?: Conceptions, causes, and assessment. *International Security* 41:72–109.

Sloan, E.C. 2012. *Modern Military Strategy. An Introduction.* Routledge, London and New York.

Smeets, M. 2017. A matter of time: On the transitory nature of cyberweapons. *Journal of Strategic Studies* 41:6–32.

Smeets, M.W.E and H. Lin. 2018. A strategic assessment of the U.S. Cyber Command vision. In *Bytes, Bombs, and Spies. The Strategic Dimensions of Offensive Cyber Operations* (pp. 81–104). Lin H. and A. Zegart (editors). Brookings Institution Press, Washington, DC.

Soetaert, K. , Petzoldt, T., and R.W. Setzer. 2018. Package 'deSolve'. Available at CRAN website and at http://desolve.r-forge.r-project.org/.

Solow, D. 2023. In models we trust — but first, validate. *American Scientist* 111:42–49.

Stanton, P. and M. Tilton. 2020. Defining and measuring cyber readiness. In *Ten Years In: Implementing Strategic Approaches to Cyberspace* (pp. 159–166). Schneider, J.G., Goldman, E.O., and M. Warner (editors). US Naval War College Newport Papers 45.

Stein, J.G. 2020. Take It off-site: World order and international institutions after COVID-19. In *COVID-19 and World Order* (pp. 259–276). Brands H. and F.J. Gavin (editors). Johns Hopkins University Press, Baltimore, MD.

Stewart, C.A., Simms, S., Plale, B., Link, M., Hancock, D.Y. and G.C. Fox. 2010. What is cyberinfrastructure? *SIGUCCS'10*, October 24–27,

2010, Norfolk, Virginia, USA. Copyright 2010 ACM 978-1-4503-0003-2/10/10.

Stokes, D.E. 1997. *Pasteur's Quadrant. Basic Science and Technological Innovation*. The Brookings Institution, Washington.

Sulyok, K. 2021. *Science and Judicial Reasoning. The Legitimacy of International Environmental Law*. Cambridge University Press, Cambridge, UK.

Szekely T. Jr, Burrage K., Mangel M., and M.B. Bonsall 2014. Stochastic dynamics of interacting haematopoietic stem cell niche lineages. *PLoS Computational Biology* 10(9):e1003794, doi:10.1371/journal.pcbi.1003794.

Taleb, N.N. 2012. *Antifragile. Things that Gains from Disorder*. Random House, New York.

Taylor, J.G. 1983. *Lanchester Models of Warfare, 2 volumes*. Military Operations Research Society, Alexandria, VA.

Thompson, E. 2022. *Escape from Model Land. How Mathematical Models Can Lead Us Astray and What We Can Do About It*. Basic Books, New York.

Thygessen, U.H. 2023. *Stochastic Differential Equations for Science and Engineering*. CRC Press. Boca Raton, FL, USA.

Tor, U. 2017. 'Cumulative deterrence' as a new paradigm for cyber deterrence. *Journal of Strategic Studies* 40:92–117.

Trachtman J.P. 2013. *The Future of International Law*. Cambridge University Press, New York.

Valeriano, B. and R.C. Maness. 2014. The dynamics of cyber conflict between rival antagonists, 2001–11. *Journal of Peace Research* 51: 347–360.

Valeriano, B. and R.C. Maness. 2015. *Cyber War Versus Cyber Realities. Cyber Conflict in the International System*. Oxford University Press, Oxford, UK.

Valeriano, B., Jensen, B. and R.C. Maness. 2018. *Cyber Strategy. The Evolving Character of Power and Coercion*. Oxford University Press, Oxford, UK.

Verizon. 2022. *DBIR. Data Breach Investigations Report*. Available at https://www.verizon.com/business/resources/reports/dbir/. Last accessed 10 April 2025.

Waddington, C.H. 1973. *OR in World War 2. Operational Research Against the U Boat*. Elek Scientific Books Ltd., London.

Warner, M. 2020. A brief history of cyber conflict. In *Ten Years In: Implementing Strategic Approaches to Cyberspace* (pp. 13–29). Schneider, J.G., Goldman, E.O., and M. Warner (editors). US Naval War College Newport Papers 45.

Washburn, A. 1981/2014 (1st edition/5th edition). *Search and Detection.* Military Applications Section, Operations Research Society of America. Arlington, Virginia (1981). 2014 edition available at Amazon.com.

Wein, L.M., Craft, D.L., and E.H. Kaplan. 2003. Emergency response to an anthrax attack. *Proceedings of the National Academy of Sciences US* 100:4346–4351.

White House. 2023. *National Cyber Security Strategy.* Available at https://bidenwhitehouse.archives.gov/oncd/national-cybersecurity-strategy/. Last accessed 10 April 2025.

White, T.J. 2020. Joint operations in cyberspace. From operational unity to shared strategic culture. In *Ten Years In: Implementing Strategic Approaches to Cyberspace.* (pp. 130–140). Schneider, J.G., Goldman, E.O., and M. Warner (editors). US Naval War College Newport Papers 45.

Wigner, E. 1970. The unreasonable effectiveness of mathematics in the natural sciences. *Communications in Pure and Applied Mathematics* 13: 1–14.

Wilber, M.Q., DeMarchi, J.A., Briggs, C.J., and S. Streipert. 2024. Rapid evolution of resistance and tolerance leads to variable host recoveries following disease-induced declines. *The American Naturalist* 203, doi:10.1086/729437.

Wouk, H. 2010. *The Language that God Speaks.* Little, Brown and Company, New York.

Zabierek, L., Bueno, F., Kennis, G., Sady-Kennedy, A., and N. Kanyeka with P. Kolbe. 2021. Toward a collaborative cyber defense and enhanced threat intelligence structure. Belfer Center for Science and International Affairs, Harvard Kennedy School. Available at: https://www.belfercenter.org/publication/toward-collaborative-cyber-defense-and-enhanced-threat-intelligence-structure. Last accessed 10 April 2025.

Zetter, K. 2014. *Countdown to Zero Day.* Broadway Books, New York.

# Index

www.ingramcontent.com/pod-product-compliance
Lightning Source LLC
Chambersburg PA
CBHW050546190326
41458CB00007B/1936